Short Stories About Power Linemen, a Dying Breed

Short Stories About Power Linemen, a Dying Breed

• • •

William H. Arnold Jr.

ISBN: 1532939949
ISBN 13: 9781532939945
Library of Congress Control Number: 2016907266
CreateSpace Independent Publishing Platform
North Charleston, South Carolina

In honor of our parents,

William H. Arnold Sr.
and
Georgiana E. Arnold

Acknowledgments

• • •

An extra thank-you for the help that my brother Woody Arnold gave me in the publishing of this book. Without his help, it would have never been in print.

Thank you, Woody!

Contents

Preface

• • •

Ideas from the Author to Help the Linemen Shortage Problem

This short book was written to share information about the job, the people who did it, the people who do it, some of the basic equipment, and how to get started if you're interested. It also tells about special people who are a brotherhood and sisterhood of very good Americans. A lineman is a worker for a utility company that climbs poles and towers to work on hi-voltage wires. America is going to have more problems in a few years as the lineman profession is dying, mainly because those in the workforce are not being replaced as fast as they are leaving.

Due to button pushers, high-tech toys, and so many people who really do not want to work, our profession is starting to have shortages. The problem will get worse as time goes on.

The people left to do this work or people becoming new workers in this field will become a lot richer due to the higher pay that linemen will be making, so there is a reason to become a lineman.

What's coming is a storm: everyone in our field will have too much work, and sooner or later, they will be tired. So keeping up the linemen force is one problem that needs attention. I believe that joining a power company or becoming a line contractor will make a fine career for a young person. A well-paying job is waiting for you. Many fun things will happen when you join groups like that because you can't be around those kind of people without having fun. My thirty-three years in the field were a blast. I hope reading that this book will influence some new people to want to join this wonderful group of people. You will not be sorry.

Shortage Information

In 2013, the *Kansas City Star* reported that there are a lot of lineman shortages in the United States. Currently, 25 percent of America's workforce averages fifty-five years old. Twenty-nine percent of workers in utility companies are fifty-five or older. Many of those leaving the workforce are linemen.

A power-engineering magazine was mentioned by Northwestern Lineman College in an information page said that by 2015, 36 percent of utility workers will be over fifty-five, and by 2020, another 16 percent may be gone from the field. That is over 50 percent of workers, many of whom will be linemen.

Now is the time to do something for our country. Companies and unions, take note. Companies and unions need to get together more, work together, and figure out something for all of us. Remember that this is for the good of the country, not just a few people.

The other advantage of being shorthanded will be more work for everyone. That in itself is a reason to become a high-voltage lineman, but you will get so much more out of it: lifelong friends, great money, and excitement (a lot of that). But another big thing is helping people. It just doesn't get any better than this last one. When I was single, I met many very nice women while working, so that is also a plus. Well, for some. The bad part is that sometimes you'll be overworked and away from your family more than normal workers are. Sometimes being a lineman is tough; many family events get canceled because of storms, call-ins, and the lineman being gone. Families have to adjust and make up activities when things get messed up because of work. Most of my friends and I had those problems. It takes a hell of a spouse to be married to a lineman. (Thank you, Karen, for being the best wife. That story will be on the Hallmark Channel one day.)

Linemen are making great salaries in 2016: thirty dollars an hour or more. Using thirty dollars an hour, time and a half would be forty-five dollars an hour, and double time would be sixty. So if you're on a sixteen-hour day, all double time, you can figure on $960 a day for one day. Of course, you will be tired, but you can't make that behind a counter at any fast-food place, and you'd be tired there also with a lot less in your pocket. Plus, you would not get the feeling of helping someone like you would after a long day of being a lineman. I am happy that now there is more money per hour for linemen. During my first ice storm as an apprentice, I was making less than five dollars an hour, double time.

The good news—for us, at least—is that many storms last many days. An ice storm in Michigan in 1985 had us out of town for twelve days. The storms in New Orleans and on the East Coast over the last few years lasted much longer. Just last weekend (April 2016) in Troy, Ohio, where I live, we had a major windstorm that blew over twenty-two ninety-foot poles in a straight line. A very lucky thing happened at that location. The poles came down and trapped cars on the road in between them. The people stayed in their cars until help arrived, which was a very smart thing to do. No one was hurt, which was very lucky. So the line crews here were on sixteen-hour days, and a lot of their pay was double time.

Just think how long it would have taken to get the power back on if only 50 percent of the linemen who were sent to the East Coast these past few years had worked. They would still be working on it, and it is April of 2016. Well, maybe it really wouldn't have been quite that bad, but you get the point: America needs linemen.

Maybe reading the stories in this book will help get some readers interested in a very rewarding and exciting job. Ask any lineman about that, and they will agree. I know that there were many exciting times during my years as a lineman.

So many fun things went on in my job. The "work family" you get will be worth all the hours, and you will make many close friends forever. Again, the close feeling and helping people are the best parts of the job. Whether on a cold day in winter or a hot day in summer, people will love you at that time for what you do, and that's a neat feeling.

There were times when no one at my job wanted to miss a day at work because no individual wanted to miss out on anything. In that way, linemen are like paratroopers on Saturday night—except they're like that every day. Most linemen are wild and fun. I know because I have been around the airborne crowd, and they are the same type of people—the best.

The first three chapters of this book are stories that were rewritten from *Power Linemen Magazine* about how I became a lineman. When I started at the power company, there were so many heroes from WWII in that group, and what a bunch of men they were. In 1961, I had the honor of meeting them and learning from them. Later, I became one, too, but not as special. They were truly America's greatest generation.

For an example, here's a story about a thank-you I received for having been a lineman; this happened twenty-two years after I retired from being one part of the greatest generation.

The Two-Star General Who Loved Linemen

My good friend General Harry C. was a flight instructor with me at the Aero Club WBAFB. That is what I did after thirty-three years at the power company. Harry had been a nineteen-year-old pilot on a B-25 bomber. He had flown more than sixty combat missions in that airplane. His copilot was killed by antiaircraft fire on one mission.

After completing his bomber tour, Harry signed up to fly P-51 fighters. He didn't come home until after the war. He was a twenty-year-old captain. I always kidded him that Goring, a commander of the German air force, found out that Harry was flying fighters and gave up. Harry always liked that. He was a great leader and friend. One night while on a check-ride flight, I was the check pilot instructor for Harry, who was a two-star general by then. After a slightly rough landing, I said, "Sir, if you land like that again, I am not saluting you when we get out of this damn thing."

Harry laughed and said, "OK, Bill. You should have been with me in '44."

I said, "I tried, but my mother wouldn't let me leave home, since I was only six." We both laughed.

About two years ago, I received a call from Harry praising a lineman in his backyard who was up a pole in a very heavy snow. We had a minor blizzard that day. He always thought that linemen did a fine job and just couldn't get over the fellow on that pole. He had mentioned before how he thought that linemen were a special breed. So after really seeing what we did in harsh weather, he called me and thanked me again for doing that work for so many years. This was from a WWII hero, so I always thought that was special. (On a sad note, my friend the general passed away about three weeks after calling me.)

That same thing happens many times from many special people. My comrade who flew in the army guard with me thought that I was something of a renegade (most linemen are, a little.) One time on active-duty training, I climbed a pole in a flight suit and hooked up a training-center trailer. We had been waiting for two days for the post engineers to hook up a three-wire service, which was a ten-minute job. So I just decided to do it. My commander asked, "Arnold, what in the hell are you doing up there?"

I said, "Sir, there will be movies tonight in the training center." He couldn't get over that.

Being a lineman has many rewards that don't always have to do with money. Of course, money always helps, and linemen are finally making great salaries. But we all received a lot of satisfaction from many other things too.

This book is written by a lineman for all linemen—past, present, and future. (It is written the way that linemen talk and about how we work, so it has not been changed by an editor. Or at least not much, I hope.)

CHAPTER 1

How I Became a Lineman— Part 1: 1961

• • •

After being an airman with no rank for four years and being let down about everything that was supposed to happen to me in the good old USAF, I received my discharge in July of 1960. Within a month I got a job as an inventory boy at a radio store. I hated that job. But I was making fifty-two dollars a week, so I had doubled my pay, since that was what I was getting in the air force every two weeks. Still, the job was boring, which made the days long, and I was working forty-four hours a week.

In the fall of 1960, I was early getting to work, and just as I was getting out of the car, I saw this guy on a pole. There was a big ball of flames near where he was. He came down the pole rapidly and was holding his arm. Soon, the emergency medical truck took him to the hospital. Since I was currently working an "inside" job that I hated working outside doing exciting work seemed appealing. I went to the power company and applied for the job of being an apprentice lineman the next day. I called the employment department every day and went there every week or two. It paid off because I got a job there after my first interview.

I started on March 6, 1961. I also added seventy-seven cents to my hourly pay and was in hog heaven: $2.12 per hour. Wow, I was rich! When I made apprentice lineman, I started climbing poles at $2.35 an hour. That was on December 10, 1961, my father's fifty-fourth birthday.

Back then, you started as a laborer and then became an apprentice lineman, and then a second-class lineman, first-class lineman, and finally a top first-class lineman. Each step in the job took about two years (except laborer). At that time, top first-class linemen were still making less than four dollars per hour. (This was around 1967, when I was a first-class lineman.) Of course, almost everyone worked overtime to make a better paycheck.

Yes, we were working with 7,200 volts off ladders and hot boards for less than four dollars an hour. We got ten cents an hour extra for working or gloving 7,200 volts hot. The most money I ever made as a leader was $13.50 per hour. I retired in 1993 as a line trouble man at that rate, and that was one reason I retired early at fifty-five. The company still had a few line trouble men doing gas leaks. Now they don't do that. I hated gas leaks, and as you'll see, I got into some bad situations doing those calls.

Linemen now make thirty to thirty-five dollars an hour (sixty to seventy dollars' double time). So that is a reason to become a lineman—in addition to helping people, which is worth something.

On March 6, 1961, my first day at the power company, I was told to get in the crew cab of this one truck, an older brown snub-nose GMC. I did, and the crew cab was black and very dark. (Just so you know, crew cabs are extra cabs built behind the front of the truck, before the back part of the truck, and they are meant for four men.) This was a laborer crew truck, and we were going to do small jobs with anchors—smaller thirty-foot poles in backyards, all dug by hand. The poles were put in with an X gin and pikes (tools used to set poles by hand). Yes, this was 1961, but we were still using 1920s tools. Now the tools are class A.

A month or so later, I was put on the Cat (bulldozer) crew, where old "Rocky Harold S." would rock the hell out of anchor holes. His truck was a newer International one with a trailer holding an old Oliver bulldozer with a boring digger for pole holes on it. It was used in the newer housing developments for new houses. This one job had over seventy-five poles on it and probably twenty-five anchors.

Once the poles were at the place where they were going to be set, Denny D., a great fellow, and Vern M. would set the mostly thirty-five- to forty-foot poles with the bulldozer, and the International truck would follow with at least two laborers, a foreman, and a truck driver. We would use four pikes to hold the poles and shovel dirt in to set the poles. Pikes were about eight to ten feet long with a spear-type pointed edge to spear the poles on each quarter in order to straighten the pole first and then hold it while putting dirt into the hole.

There was also an air compressor and hoses for two air tamps. It was a pretty good system, and one day I remember that we set eighteen poles, which was really good. Let me give you a little more information about that time. The foreman really liked to drink; he'd have a double-double with coffee and an egg sandwich during coffee breaks when we stopped. This was pretty much every morning. We liked him needing a drink because we would get a coffee break. When he didn't need a drink, we would go straight to the job. It was always a harder day when we did that.

Denny drank beer all the time; it killed him within eighteen months. He was killed in a 1952 Chevy convertible one Saturday night on a dark country road. He drank too much but was a hard worker and good man.

Thinking back to this time, I remember the two other digger trucks. One was a 1927 truck. Remember, this was 1961, so that was a pretty old truck. It had two seats in front and a crew cab attached for the backseat. Three or four people could fit in the back. The other digger truck was smaller with only two seats. On town jobs, we would use the International truck to follow the diggers. Sometimes we would get to be around a line crew. The first time, we needed the only bucket truck that DPL had in Dayton to cover some primary (primary wires are 7,200 volts each). Covering primary meant putting rubber hoses about five feet long on the wire in case the new pole touched it while we were installing it. If that happened then, we were still safe. If the new pole touched the bare wire, we were not.

That is when I met Duke C., a first-class lineman (there are a lot of stories about him in this book). I know one thing: whatever he wanted, I wanted. He was a great fellow, a great lineman, and a great American. Like us, right? As an apprentice lineman, I was issued tools, a belt, gloves, and a hard hat. We had to buy our own climbing hooks. My first pair was Brooks hooks. Later, I bought aluminum left and right Bashlin hooks and had them for thirty-two years. Our climbing training was done at dinnertime or during break time. Sometimes at night, I'd go out to a private road and climb bare poles up about twenty feet.

After being an apprentice for a month or so and learning how to make up the hardware to connect the wires to. We started climbing thirty-five- to forty-foot poles. I'd climb one hand for me and one hand for the company for a while. After a month or so, we started doing better, and we got a lot better as time went on. That was the only climbing schools they had for us at the time. We went to line classes at night on our own time. Now you go to linemen training schools first to learn all of this (there is information on these schools later in the book).

As a-year-and-a-half apprentice linemen, we were put on the streetlight truck. This was an old box-lift truck where you had to crank the bed around to where you needed to work. It was a crap truck. The thing had a telescope ladder on the side.

I have to jump ahead here. As a new leader in 1979, I was on this same truck. I was up on the lift to help a new fellow, and the truck froze up. We couldn't move anything. There was no one on the ground to help us, there was no radio in the bed, and there were no cell phones. So I climbed down the telescope ladder. Just as I got off and was clear on the ground, it collapsed about five feet. I almost lost my hands and feet. I never took that truck out again. I put a ladder up to get the new man off the bed. Again, linemen have class-A tools now.

The company started getting corner-mount trucks in the mid-sixties. They made up a new job using a crew leader with a first-class lineman and a second-class lineman on these crews. There were a lot of small jobs, and they were a better truck than the old digger trucks. When I started, everything at that time was done on a GMC snub-nose truck. The trouble trucks were starting to receive the pelican single bucket for trucks. As a new second-class lineman, I was on a truck number 629. (For more, see one of my stories in my blog about pilots and linemen: http://linemanaviator.wordpress.com/. Check out the story about the new second-class lineman, or November 22, 1963. It was a hell of a week for me and our country.)

As we received more experience on the trouble trucks and became better linemen, after about six months, we went over to the service department to hang services. This involved 120-/240-volt wires coming from poles to houses. I was there for about four months. Then I went on the leader truck. This was where I started working on primary or 7,200-volt wires, hot.

My leader was a WWII navy veteran named John C. He had been a sailor on those flat-bottom landing crafts that opened up and unloaded tanks and ducks. (Ducks are boats for hauling troops.) His was a very hard-riding ship. John said it was a bitch in a storm. He was in the famous typhoon right after the war. He said it was the worst thing he ever did. When I started at DP&L, there were many heroes from WWII in the line department.

We were on #1073, a GMC flat-nose corner mount. John sat in a chair on this one. It was one of the first we had. Later, the operator (leader) stood in the left back of the truck to operate the boom and digger.

The time period was fall of 1964. A first-class lineman and a second-class lineman were allowed to do small primary jobs; these were high-voltage, switches, transformers, and single-phase type of jobs. This was when the company started selling night guards (small lights). We did a lot of those.

My first big ice storm was with Big John and Harold J. in late February of 1965. Of course, we had to climb everything with two safety belts. On our way toward the storm, about ninety miles from Dayton, we heard a mayday call on the radio. This was the first time for me to hear that. A fellow fell forty-five feet as he was only using a single safety belt. He unbelted to go over a secondary bracket on an icy pole. That was a no-no in those days and even now. You never unstrap on an icy pole. But there is a new way that linemen climb poles now. It's a lot safer. Bucket trucks get almost all poles now. Yes, you will still climb poles, but less than half of what the older linemen did. I know one lineman who is now retired and didn't climb any poles for the last fifteen years that he was with the company.

In December 1965, I made first-class lineman. We still had only one bucket truck for big jobs. Most big "hot crews" had two first-class linemen, a second-class lineman, a driver up front, and a foreman. I say "up front" because we rode in a crew cab behind the foreman and driver. Flat-nose GMCs with a crew cab were a little nicer than the first truck I was on. Most trucks had a small table in the crew cabs for eating lunch and playing cards when it was raining. They were all made by the men on the truck and were usually supported on each end by #6 copper wire. At this time, we were still earning under four dollars an hour for pay. From 1965 until 1972, I was on three crews (we called them "hot crews"). Hot crews were in a GMC truck, a trailer with whatever we needed for the jobs. The clothes we wore were mostly Levi's or jeans. It was kind of neat in the fall when it started getting colder. All of these colorful flannel shirts started showing up on the dock in the morning. JC Penney's had great flannel shirts for three dollars. We didn't have uniforms until later—way later. Of course, most fellows wore Carhartt's bib overalls with a jacket or a hooded sweatshirt as it got colder. (There's another story about that in the blog.)

We were still climbing everything on the streets. We used hot ladders a lot; those were eight-foot ladders with large hooks on them to hang over cross arms. Or we used the hot board for smaller jobs and smaller poles. We used both boards and hot ladders for working twelve-thousand-volt wires hot. They isolated us from the pole, which was 90 percent ground. Grounds were our enemy. While cutting in airbrakes (big switches), we were on each side of the pole on hot ladders.

Linemen become close, as do soldiers and police officers. We kept each other alive. Any linemen reading this know what I am talking about. Each man would watch behind his pole buddy to help keep him safe. Working like that was our combat.

One time, a short lineman in Florida told me that if you put every lineman in a bag, shook it up, and dumped out a couple out, you couldn't tell any difference; they were all the same. That's pretty much true. I have always thought that being a lineman was neat, and not everyone could do it. I still do think that. We worked hard, but we had a blast.

CHAPTER 2

How I Became a Lineman—Part 2: 1970–1982

• • •

As promised, here is some info on Duke C. An article in *Power Linemen* magazine mentions Slim, a lineman in the 1930s played by Henry Fonda. Preston Foster, another old actor, was his side-kick (that's actually a mistake, but I'm going to leave it and explain later). I have seen that movie, and Duke C. was our Slim.

Most of the early linemen came from contractors, like Duke did. Here's another bit of information from the time period: Mickey Mouse insulators were called "donkey ears" until the Mickey Mouse movies came out. After that, they were called "Mickey Mouse" insulators. A foreman told me that when I first came across them.

Here's a picture of one to give you an idea of what they look like. Very pretty.

My first impression of the men that I met on my first day of work was "When I grow up, I want to be just like these guys." They all seemed happy and cheerful, and they had a bond between them. I later found out why: because linemen are like combat soldiers, keeping each other alive when they're working. Now it is so much safer with the buckets, but back then we couldn't help but become like brothers. We would even watch out for guys we didn't get along with. I wanted to be in that group. And I made it—for almost thirty-three years. We are still close, as I am sure all linemen are. Well, most are anyway.

Duke C. was a first-class lineman when I got there. Wow! He was a mild-mannered fellow, but he could fight. He stuttered and called everyone "l-l-little man," as he was a well-built guy.

In 1962, I had my first bucket ride with Duke on truck 1016. It was the only bucket truck that the Dayton branch had at the time. He took me up, and I touched a twelve-thousand-volt line for the first time. When I touched it with my pliers, it spits an arc at my pliers, and I jerked my arm back.

Remember, Duke stuttered when he talked. He laughed and said, "T-touch it again, l-l-little man. It's t-talking to you."

I asked, "What's it saying, Duke?"

He said, "'I am going t-t-to get you. I am going t-to get you,'" and he laughed. We did our work and went down. About nine years later, it got him.

Duke first came into town in the early '40s with a bag of clothes and a bag of tools. He was the last of the roving linemen. These guys would work their way across the country from job to job. Duke worked for the bus company first. I knew a guy from there, Fish, who had worked with Duke. He then came to the power company.

Duke lived in rooming houses and would move every six to eight months. I do not know why. He came into work one morning after WWII started and told the boss, "I won't be in next week."

The boss asked why, and Duke said, "Because I joined the navy."

The boss asked, "When are you leaving?"

Duke said, "As soon as I get off work." Most people would take a few days off. Not Duke.

He was a great guy, but I guess he got into a lot of trouble in the navy, as he kind of had a mind of his own. I do know that he had a choice: the brig or UDT (underwater demolition team), so he ended up as a frog man. I guess he saw a lot of action. He never would talk about it.

Most linemen in their thirties in 1961 were WWII veterans. We had a lot of heroes in our company, many of whom were paratroopers.

Another man who would not talk about his service was my friend Wilson. He had been a gunner on a B-17 in Europe. He told me one day that he just could not talk about that time of his life. I was trying to get him to talk about the B-17. I never did again.

Most of those men did not talk about the war. Wilson had bailed out of two B-17s on fire and had twice fallen through formations of B-17s and made it to the ground. Now, that would scare anyone. The first time he made it back to England and was right back in the air. Another time, he was a POW for over a year. Wilson passed away two years ago. He truly was a silent hero, like most of the WWII veterans.

There were a lot of stories about Duke. He was our Paul Bunyan. In the old days, line crews would bust their butts during the week, but on Fridays, they would work a long morning and quit for the rest of the day, cash their checks at a bar, and stay there until quitting time. When I first got to the power company in 1961, I was shocked at how much people drank while working. That has changed and doesn't happen now.

One Friday at about 1:00 p.m., Duke finished his pole, came down, put his tools away, and went into a bar across the street. About ten minutes later, he was in a fight, throwing two guys out of bar. The other fellows from work were just putting their tools away and rushed into the bar. There was Duke at the end of the bar, sipping a beer. The guys all rushed to Duke and asked, "What's going on?" Duke said, "Nothing. I f-f-felt like d-drinking by myself, l-l-little man." You can see why no one wanted to miss a day at work. You never knew what might come up. It is still that way. What a group!

In 1969, about eight years after I became a lineman, we were still on the hot crews climbing 99 percent of everything. The crews had a line truck and trailer with two first-class linemen, a second-class lineman, a driver, and a foreman. In the Dayton line department, there were about six of these crews. By then, there was another double bucket truck. So now we had two. Somewhere in this timeframe, the company started having service centers. So they started sending linemen and all jobs to these centers.

I had been working with a great fellow named Jim H. (my pole buddy) for over two years, and we were still climbing poles right on the street, doing big jobs off ladders and hot boards. Now we'd be in a bucket on these types of jobs. In 1970, we had a big job just north of the city of Dayton; we were probably on it for five months. The service centers seemed to have smaller crews for the leader crews, and thinking that I might not have to work so hard, I transferred to the Salem service center, which was on the northwest side of Dayton. Well, just as I got to the new center, thinking I'd be put on a smaller crew for an easier day, they decided to have another big crew, and guess who got to go on that big crew? Right. Me. I worked on it with my friend Deo, a former US Marine and a very good fellow. (He still is.) We worked for over eighteen months together on that hot crew, still climbing poles all winter on a highway right by the Dayton International Airport. It was a very cold and windy winter.

So now we were getting more buckets, and it sure helped the knees. In 1971, I had just become a commercial pilot with a multi-engine rating. I wanted to be DP&L's air force, so I learned to fly helicopters. In 1972, the company decided to start a high line department working 345,000 volts hot off towers, where they would probably get a helicopter, as one vice president of the company told me. So I applied and received one of the five transmission lineman jobs. I had been at DP&L for eleven years. The day I put in for the transmission lineman job, my father passed away. It was a sad time.

By then, each service center had a double bucket truck or two smaller buckets on at least one crew per center. When we started using buckets for many things, our safety record got much better. Yours will be good, too, as you'll be on a bucket truck.

The transmission department had a ninety-foot bucket, which was great on ninety-foot poles. But there was still climbing as we couldn't always get the truck to the site. The other bit of news was that the transmission department worked out of the Dayton service center. So I was back in Dayton. In 1972, we were making eight dollars an hour, which was pretty good at the time.

When we worked 345,000 volts hot on 150-foot towers, it was tough. We climbed the towers and hung a twenty-foot fiberglass ladder over the side. We all wore special boots while doing tower work. They had steel shavings in the soles and a wire coming out of them to attach to metal plates that we would put on both bare legs. We were grounded to save us from being zapped by static electricity. Steel towers are all grounded; when we stepped over onto the twenty-foot ladder to climb down, we were no longer grounded. We couldn't go up that high and be around all of that static without them. We had tried it without the boots and couldn't go past the wire in insulators. That was when it really hurt to try it.

Here's a typical day doing a job on a 345,000-volt wire. Two men would climb down the ladder and stick the wire with seven-foot fiberglass tools or sticks. That 345,000-volt wire had about an eighteen-inch field around it, and when we put a stick or tool in it, it would arc and burp loudly. And I mean loud. I always pulled the stick back when that happened, which usually made us laugh, and then I'd go ahead and do the work. It was very loud the whole time the stick was on the wire while we were sticking wire off a ladder. When we were done, we had to climb back up ten to fifteen feet and rebelt, two rungs below the tower. We took off our leather gloves and bare handed the grounded tower. We got shocked by bleeding our bodies of static electricity as we had been down in that field of electricity. It was like sticking your finger in a power plug in your house for a moment. If you let go of the tower, you got it again. If you touched your pole buddy while down on that ladder, you would also get shocked as your two bodies were equalizing the static electricity in both of you to be the same. When linemen work it hot now, they mostly do it with helicopters.

In 1979, after seven years as a transmission lineman—working towers, ninety-foot poles, and 345,000 energized—for fifteen cents more an hour, I made leader lineman. Then in 1982, the company had a big layoff, and I got bumped out of the leader job. So I bumped a line trouble man in Fairborn, Ohio. It was the job I had always wanted. By then I knew that I wasn't going to fly for DP&L, so the bump worked out well for me.

I was now in Fairborn. In that town, line trouble men would go out with a leader crew and do everything. If there was a trouble call, I'd go do the trouble call and then return to the crew. In 1986, the company wanted a line trouble truck working from 4:00 p.m. to midnight. They wanted every lineman in each center to take a week at a time to work nights. In Fairborn, no one wanted to do that except me. One big reason I wanted to do it was to be able to run my mother to the doctor in Cincinnati, Ohio. To do that, I was taking off without pay two times a month. So the other linemen stayed on days, and I went on nights' full time. I would report to Fairborn for a four-to-twelve shift. I had a truck ready to go. It was an All-tech bucket (8451), a great truck.

Of course, all overtime after midnight was double time, and all call-ins were double time, so it was a good thing for me. Plus, when taking my mother to the doctor, I didn't lose pay. I loved trouble work and did a good job at it. It was very exciting.

It also let me meet the night people. There is a different bunch of people out at night—not all bad, just looser. There were some bad people, but most were a good bunch. I also knew all the police and firemen. Fires and wires—I loved the trouble truck. I liked being alone as it let me make my own decisions. That was why I was glad to be on nights.

Then, in 1990, the company started a trouble truck (emergency department) working out of another center in Huber Heights, Ohio, about ten miles northeast of Dayton. I worked from four to midnight Sundays through Thursdays. This is when I started going everywhere in about a fifty-square-mile area. My mileage started really going up. It was because I was going to about six service center areas. The shifts went fast as we were busy.

I got in trouble asking another lineman on company radio if he wanted to use my truck on a very snowy night; I had been on for eight hours, and my truck was warm, ready to go, and fully equipped. The slab head boss (see the next chapter) said I was rude to him, as we had been told to use our own trucks. He was not a lineperson. That's enough on that subject—except someday I hope to see him in a parking lot.

CHAPTER 3

How I Became a Lineman—Part 3: 1982–1993

• • •

"SLABHEAD" WAS ONE OF OUR nicknames for anyone who didn't do something right. A slabhead was somebody in traffic who did something dumb. Or if you dropped something, you were another slabhead. It was started in around 1972, when I became a transmission lineman, by Bill T., the leader who bumped me to line trouble man. He was the Tom Sawyer of the department—a hell of a man. You could paint his fence, but you'd have to pay him to do it, just like Tom Sawyer.

I was a slabhead in the last chapter when writing about the movie *Slim*. Pat O'Brien was Slim's buddy in that movie, not Preston Foster. I made a mistake. But like I told my flying students for twenty-five years, "I don't care if you make a mistake; just don't kill me."

Anyway, I loved the trouble truck. I liked being alone or with a new helper. Around 1992, the linemen were being hired as second-class linemen. They started coming from training schools. I always had good helpers from these schools, mostly cowboys or farmers. All of them at that time were from out West. They all had good work ethic. One of my young helpers from 1992, who is now a transmission leader, told me recently that soon he is going to let me ride along and check out new things. As I said, the trucks are much better now, and I am looking forward to that. Due to DP&L's insurance policy, I haven't gotten to do that yet.

DP&L started making us wear uniforms in the mid-eighties. The first ones were a light-brown shirt and dark-brown pants. We had Carhartt jackets and dark-brown hooded sweatshirts with bib overalls for colder days. It saved money, and I didn't mind at all. Now they wear blue uniforms, and the clothes really have changed a lot. Everything has. It's nicer and safer, with green rain suits that glow in the dark and rain jackets that probably keep people drier than the old, cheaper rain suits we wore from 1961 to 1993. They are also safer for the workers since you can see them easier in backyards when the power is out. People know the look of the rain suits with the white hard hats, which is much safer.

We were still on the twelve-hour day shifts, seven days straight, three days off, seven more on, four days off, seven more on, and then a week off or something like this. Being off for a week was like a mini vacation.

During my last year at the power company, I had mostly twelve-hour day shifts, and I didn't mind. I was married, and I actually loved the time off with this harder schedule.

My last helper on the trouble truck was a small woman (DP&L's first female lineperson) who was around five foot two. She was an apprentice lineperson. Not thinking, I had her cut down a long duplex streetlight service, as I thought she must have done line work longer to have gotten on a trouble truck. When she did, she tried to hold the wire, and it damn near pulled her out of the bucket. I had to yell, "Drop it!" After that, I worked a lot with her, trying to teach her to let the equipment do the lifting and not her. She was too small. I even worked with her on my own time to try to help her.

To pass one test, she had to hang a double nine-foot cross arm in the hooks on a forty-foot pole with a single hand line. Before our shift a couple of times, I worked with her for a few hours to show her the easiest way to do that. She did pass that one and went on for a while longer after I left but had to quit as she got banged up. So for any new people coming aboard to do this work, here's your first lesson: You cannot bull this equipment; you have to learn how to handle things so that you do not hurt yourself. Everyone becomes a rigger or someone who rigs things to let people on the ground help hold them.

On August 1, 1993, one day after turning fifty-five, I retired in my thirty-third year at DP&L. My last call was a gas call. The trucks were really nicer and safer, but we still looked up addresses with the dumb book and only had a radio. Now they have GPS, computers, and printers—wow, what a change. Why, they even have air conditioning!

I have always been proud of having been a lineman, and I still think a lot of all our brothers and sisters in the field. I have always believed that not everyone can do it, and I always will. God bless all of you and those who follow us.

CHAPTER 4

One Night on the Emergency Shift: 1991

• • •

I WAS A LINE TROUBLE man for my last ten years at the power company. I worked mostly nights, normally alone. I had all kinds of calls: poles hit by cars, wires down, and blown fuses to major sections out of electric. I also had quite a few people killed and hurt from having contact with our wires. Seventy-two hundred volts will kill anyone, so when you see a wire lying on the ground, stay away from it. Never touch it, and keep other people away. Get the power company there ASAP. I've seen wires do everything; you never know if they are energized or not, so never touch them.

It was a fast-paced shift: four to midnight, Sunday to Thursday nights. This was my favorite shift. It was very exciting. On Sunday nights at midnight, I always figured that my Monday was over. Most of the calls were on Dayton's west side, which was the older part of the city.

This one Sunday night about 7:00 p.m., I received a call at a funeral home in the west end of town, just west of Interstate 75 and about two blocks east of Broadway Street. They were rebuilding the whole area, and this funeral home was about the only thing left that wasn't being torn down for rebuilding. The call was an inside problem or the customer's problem to fix. This funeral director was out to get the power company to fix his problem, so I had to check everything to prove to him that it was his problem.

After I had checked many things, he led me through a room with a dead woman and many, many mourners. Now, these fine people were really getting on with this mourning. I mean the funeral home had nurses in white dresses with smelling salts and aspirin. I have never seen people mourn that loudly or in that manner before.

I walked past the casket and into a back room. After checking everything, I decided to check the wall socket behind the casket with the lid up and the lady's body in it. So as not to bother anyone, I got on my hands and knees to get there, and no one saw me, but the bad part was when I was done. I was thinking about going to the funeral director and how I would be telling him that it was his problem.

Not thinking, I stood up. There I was, six foot four with a hard hat on. Now, remember, I was behind this casket where the body was, and the hall was packed with mostly women, and the screams were the loudest that I have ever heard. I scared the heck out of those people but not on purpose. Actually, I can understand why: all of a sudden, this tall guy in a hard hat shows up behind the body. So the smelling salts were used quite a bit that night. I ended up leaving, and the customer wasn't too happy because it was his problem to fix.

The next call was a guy stealing electric from the power company. He had a massive short and got his house to catch fire. When I got there, the fire department was already on the scene. After seeing the jumpers (wires bypassing the meter) around the meter, I had the dispatcher call fraud and theft from the power company and also told him to send for the police. Fraud and theft was a department in the power company that looked into jobs where people were stealing electricity. So, overall, it was a fun night.

All said and done, the customer was handcuffed, his house had major damage, he was in trouble, and he was on the way to jail—all for a few dollars off of his electric bill. It's not worth the effort to do things like that. The power company will always find out what is going on.

The next call was a gas leak—yes, I said a gas leak. They had linemen doing gas leaks at this time. I hated those calls because I felt I wasn't trained as well as I should have been. Plus, they were always dirty calls. So I got into this dirty basement and was looking around, and I found the problem. The customer had been trying to relight the water heater and called it in as a gas leak. I relit the water heater, and as I was doing this, the cage beside the water heater kept making a loud thudding sound. After lighting the water heater, I pointed my trouble light on the cage, and there was a boa constrictor snake in the damn cage lunging at me and hitting the side of the cage. It was time to dust my broom, which means to leave. I would have rather been climbing a pole on fire than doing gas leaks at any time. I had dispatch mark a comment about a snake at this address for anyone else going there.

The rest of that night was just a nice, peaceful fire call, cutting down the wires so that firemen could enter the building safely. And that was one night on the emergency truck shift in 1991.

A new note on boa constrictor snakes. A year before, I had been on a fire call in a building with the firemen making the electric safe for them. This fireman came down the burning stairs as we got out of the building and said that he went into one room, checking for people, and there was a boa constrictor snake loose in the room. I asked, "What did you do?" He said, "I slammed the damn door shut and let that room burn a little more." Later, we found out that the snake had died.

I wonder if most linemen are like me: as you go by a pole or poles, you can remember when you worked there. I get kidded by my non-linemen friends for talking about the job as we drive by places where I have worked.

I loved being a lineman and would have stayed for five more years if it hadn't been for the gas calls.

CHAPTER 5

The Day We Blew Up the Power Truck

• • •

IN THE SUMMER OF 1961, I had just turned twenty-three years old, a laborer waiting to become an apprentice lineman at the power company. Laborers did all jobs: setting poles and anchors, plus all kinds of work getting jobs ready to build a line. The company had an old power truck from WWII. It had an air compressor, and we broke up cement with a jackhammer or drilled holes into rocks with an air drill to put dynamite in to blow up rocks. Union Ohio has a rock base down about four feet, so we worked there a lot. For example, a forty-foot pole is set about seven feet. So we would have to drill three more feet into rock, blow it up, and clean out the rock for the seven-foot hole.

Normally, there were two laborers and an older, leader-type fellow on this truck. On Friday, the leader wasn't there, and the two laborers were sent out with one truck. Neither of us was trained. Just this same week, Freddie and I had been on the crew all week, and DP&L had an ex-foreman doing the dynamite work to show us how. That was the training we received for about five or six holes. The ex-foreman was not there on Friday either.

So off we went with the truck with no radio but with dynamite and blasting caps. That was a no-no, we learned later. We had no business doing that job. The morning went well, and, after lunch, we were working hard to get this one hole done, but it was getting late.

We wanted to get a Coke on the way in. Normally, we drilled four holes and put one stick of dynamite in each. I talked Freddie into putting two sticks in each hole so that we could get the job done and get our Cokes. Freddie said that would blow the steel mat, which we put over the hole so as not to throw broken rock everywhere, into the sky.

I told him, "Let's pull the truck over the hole to cover up the mat and to keep it from going airborne." Long story short, that's what we did. Getting ready to hit the old type of plunger that you see in movies, Freddie said, "I don't know, Bill." I said, "Get down, Freddie" and pushed down the plunger that fed eight sticks of dynamite into a five-foot hole with the steel mat and power truck sitting over everything.

Well, this truck goes straight up about a foot, and it comes crashing down, rocking side to side. It almost hovered for a minute and then came down. It actually was kind of funny; you had to be there. When we went to check everything, the hole was deep enough, and the truck was OK except for a little dent in the muffler. I said, "Let's go get a Coke, Freddie."

Another thing I remember is that, at lunchtime, we were in Englewood, Ohio, and there were many construction trucks parked on the street. Our truck had that dynamite in the back, and no one ever touched it. It was a different time. I don't even think that anyone even thought about it. Darn, I miss those days.

CHAPTER 6

The New Second-Class Lineman

• • •

I HAD A TWO-YEAR APPRENTICESHIP and was going to night school; it was like being a new soldier in boot camp. It was pretty tough, and you really had to want to become a lineman to get through all the hazing and stuff we had to do. Of course, the older apprentices and the foreman gave us a hard time. But just at the two-year mark, I made second-class lineman. That was in early November 1963. It meant about fifty cents more an hour at the time.

It also meant that I would be going on the emergency trouble trucks to pick up more experience in working hot wires and everything. The trouble trucks were two-man trucks with a line trouble man as the boss and the second-class person as a helper. It was a pleasant change, not having a foreman on your ass. It was also a great way to learn more about line work as we talked over each call that we had been on. It was a job that worked first and second shifts at the time. Of course, working late on second shift gave us double-time pay, and we all liked that. All overtime after midnight was double time, and that really helped a lot—especially as I had been working for three years at lower pay.

The top pay in the US Air Force was just fifty-two dollars every two weeks, and there was low pay for the apprentices too (I started climbing poles at $2.35 per hour). So now I was starting to make more money so that my family wouldn't have to struggle as badly. I probably made around three dollars an hour. Plus, I was starting to work more and get more overtime, which really helped. We called overtime pay "gravy money," and it was a nice change at the time. (When I became a top first-class lineman, which meant I climbed everything and worked twelve thousand volts energized every day, I was still not getting four dollars an hour; I believe it was $3.95 an hour.)

My first night on the trouble truck was November 17, 1963, a Sunday. I was assigned to work from four to midnight with LR, a great fellow who had been a combat soldier in Korea and had seen a lot. He had been off for a while as his helper in the summer had been killed (this same fellow is in the "Your Side" chapter). LR was very upset about Jim's death, plus as I said, he had been in Korea in some very bad things. He was a quiet, soft-spoken, very kind man.

We went through Sunday, Monday, and Tuesday OK, and then came Wednesday. About halfway through the shift, we got a call somewhere on the east side. It was a battle between a squirrel and a transformer. The squirrel lost. The transformer had an old-style switch with a fuse in it. I got it out and sent it down to LR, and he refused it. When I say "down," I mean I was up in the bucket about thirty-five feet. We used a fifty-foot rope to pull things up and to let things down with. This was in truck 629, the first trouble truck with a bucket. I put the fuse in the switch and closed the switch with a hot stick. All the lights went back on.

About an hour later, LR thought he had put too big a fuse in the fuse holder. So we went back to the job. He wanted me to go up in the hot wires and put a wire jumper around the hot switch to keep it working while I refused the fuse holder. By then, it was night and quite dark. I was a brand-new second-class lineman with no hot-work experience with jumpers or anything. So I told LR that I didn't think I should do that job because of my lack of experience. Long story short, LR did the work, and when he came down, he was upset about something. I told him I would get the truck ready to travel and for him to sit down. I ended up driving the rest of the night. LR kept saying, "A man ought to know better than to work that stuff" or something very close to that. He kept repeating that for the rest of the night until we went home.

The next night, when I got to work, the supervisor of the emergency crews called me into his cubicle. Bill P. was a Battle of the Bulge soldier who almost lost his feet due to frostbite. He always had trouble with his feet all his life. Anyway, he wanted to know what happened the night before, and I told him. LR ended up being off for quite a while and switched jobs to installing 120/240-volt service wire into homes.

LR was a great fellow and an old friend. What he went through in Korea, what happened to his helper while he was on the ground (he had to get Jim out of the bucket by himself), and everything else he went through that day was all very hard on the man. People are more understanding on this type of stuff today. (God bless you, LR.)

On Thursday night during my first week on the trouble truck, I went with an upgraded first-class lineman called Louie M., another soldier. We made it OK, and on the way home, I bought doughnuts for a little surprise. Of course, the next day was November 22, 1963. It was my first day off from the trouble-truck shift. But it was a day to remember for our country.

That was the day that John F. Kennedy was killed. The world really did change that day. (My first son, Scott, was four months and seven days old then.) This was the first time something like that happened in the United States, and Walter Cronkite really took over. The whole country was glued to the TV all weekend. On Sunday night, I watched as Oswald, Kennedy's killer, was shot by Jack Ruby. What a night that was. It was the first time that news people took over the networks. Actually, they did a good job as they reported the news. When Cronkite told us about Kennedy's death, he was shaken like all Americans were. It was a sad time for our country.

The following week, America had a lot of TV news shows with the funeral, parades, and the lighting of the eternal flame for John F. Kennedy. We worked a normal shift on that Monday, and that night, a small John-John, President Kennedy's son, was on the news saluting JFK as his body went by in the funeral parade. Later, John-John would die in a plane crash, which was something else that didn't have to happen.

That was my first week as a second-class lineman, and it was a hell of a week for all of us and our country.

CHAPTER 7

Ice Storm: 1965

• • •

THIS CHAPTER IS ABOUT MY first ice storm, where we actually climbed the ice. We climbed the icy side of the pole so that our safety belts would be on the dry side of the pole and wouldn't slip. When we climbed up to something, we would never unbelt. We had another safety belt to put over whatever we were climbing by. Sometimes it took a long time to get up poles. On this first job, these were ninety-foot poles. I have never been more scared in my life than I was on that day. It was windy, with blowing snow, ice, and cold.

In the month of February 1965, I was on a three-man crew (1073) with the leader, John C., a sailor from WWII. He had come up the lineman-training job progression, working for the contractors. At the end of WWII in the Pacific, John had been on a flat-bottom boat that would open up in the front to let out tanks and troops in smaller boats. He was in that famous typhoon that really scared all of the sailors. He told me that was something he never wanted to do again.

When I started in 1961, "Big John" had been a line trouble man (my dream job). I was a second-class lineman by now (1965), and Harold J. was a first-class lineman on this crew. I had just spent six months on the trouble truck as a helper, and this was a step up to learning more line work by working hot wires.

The last week of February 1965 produced a bad ice storm up north around Grand Lake. There were many outages (people without power). In the early morning, they started getting crews together to go north. The weather was nasty with ice and snow coming down hard. After loading up and packing a bag, Big John, Harold, and I started north. Of course, I had to sit in the middle; Harold was the first-class lineman, and second-class linemen didn't sit by the window. Big John smoked these Petr cigars from Cuba, and they smelled like crap. He would always flip his ashes on me, sitting in the middle, and then laugh. It took us a long time to get up there. On the way, we heard someone screaming, "Mayday, mayday" on the radio.

We didn't like hearing that. In the beginning, I was excited—and so were John and Harold—because we would make more money and go somewhere, but now things were getting a little more complicated. We stopped on the highway to talk to another leader, Art B. He had Jerry D. and TC on his crew. Jerry was up to the top of a ninety-foot pole, and one side of his body was covered in snow and ice. I'll never forget that image. It was a harsh day, and there was still ice and snow coming down. Not only that, but the pole was full of ice, and Jerry had to use two safety belts to climb the damn thing. It was very hard, tiring, and dangerous.

When we talked to Art, we found out that a guy had fallen off an icy pole and had been killed. On icy poles, you never climb the damn things without two safety belts, and you never unhook a safety strap. He only had one safety.

It took a lot of energy and strength to just go up one of those things. We went on into the service building to pick up supplies. We were going to a transmission line (ninety-foot poles) to change cross arms that were broken. While loading the supplies, Big John fell out of the back of the truck that was full of ice and landed right on his back. That would have stopped a normal person but not Big John. After a few minutes, he went about his business of loading the icy truck and finding out where we had to go.

We found out. Damn, did we. When we got to our job site, it was dark and still blowing snow—say, gusty forty-five-mile-an-hour winds. There was a full moon (a bombers' moon, as the Brits would say). It was a good thing, as all the electric was out everywhere, so that was our best light. We could see by the moon hitting the snow on the ground better than we could see with anything else. The pole we had to fix was over a mile down this railroad track. We had to carry all of our tools, hooks (what we climbed the poles with), and linemen belts, plus big rope blocks, trouble lights, two nine-foot timbers, six insulators, and a few small ropes called slings as well as a double hand line to get the stuff up the pole to us. This was a night that was just like soldiers experience a lot, like going on patrol. You couldn't help but wonder how this would end. We knew it wouldn't be good where we were going. And it wasn't.

It took quite a while to walk down this railroad track. We wore our tools because we had so much stuff to carry. Blowing snow and ice were all around. When we got to where the pole was, we had to cross a ditch, filled to the brim with snow, to the fence line and climb a fence. Going over that ditch (we used the pole cross arms to cross the ditch), I slipped off into four feet of deep snow. Harold and BJ got me out of it. Then we had to climb over the barbed-wire fence to get to the fixtures we were going to replace. It was for a sixty-nine-thousand-volt line. There were timber cross arms and four bells for each wire. Harold went up first. He made it ninety feet up. He found that there were two broken timber nine-foot cross arms that we had to replace. Timber cross arms are bigger than normal nine-foot cross arms.

Now it was my turn. Damn, I was scared. Also, I had to carry up the hand line as I was the bottom lineman. This was the double rope we used to have things pulled up to us. It was heavy too. I started going up, and about forty feet up, my left hook came apart at the adjustment and made me miss the pole properly. I fell, sliding down that icy SOB, and came to a halt about twenty feet up. I got hooked back in. (Hooks were the tools on our legs that we climbed with.) My hooks came apart because of walking down the railroad tracks with them on. The screw that set the adjustment for length came out. I had two screws in the right hook. I took one out and put it into the left hook. Then I climbed that SOB and made it to the top with Harold.

We could just barely see the light from the trouble light up there. It was still blowing ice and snow. I said, "I am scared."

Harold said, "Don't feel pregnant."

We were both scared. We got those cross arms changed in pretty good time and got the new insulators changed too. While I was up there, my hard hat blew away, and I never saw it again. After two hours, we made it down and got the line back on, picking up a big section as the voltage on that line was sixty-nine thousand volts. We had to walk a mile back out also carrying a lot of tools. We left everything else for crews in the spring to pick up. It was now very early in the morning.

We were tired and wanted some coffee. Keep in mind that this was 1965, and there weren't Speedways or many places that were open all night. On the way to get some coffee, we saw this big blue arc in the sky and knew what it was right away: a wire burning down. We saw it fall. There went the coffee!

We had to fix it. We did another icy pole, but at least it wasn't ninety feet. We worked through the morning until around 11:00

a.m. BJ was very sore from his fall. We got five hours of rest and a meal. After five hours' rest in an old motel without heat, out we went again. We worked for a while on services, 120-/240-volt wires going into houses. Then the storm was over. They put us up in a flea-bag motel that had a bar. Did we sleep? Hell, no! We drank. In fact, I have never been drunker. (Note: I have not had a screwdriver since that night.)

It was all linemen at that party. We sang, talked, and got wild. It was fun. I felt like an airborne ranger. And I loved the men I was with. We had just come through a hard time and were still alive. Not all of us would be in a few more years.

In those days, the DP&L line department was just like the military. We had real men running the outfit, not like some of the people who have had the same jobs in the last few years. When I got home the next day, I got tears in my eyes when my first son, Scott, came to the door. He was nineteen months old at the time. I think it was because I had been so scared. Or maybe it was because my head hurt so badly (ha!).

It actually was a short storm. Later, we would be gone for days before being released. But this was my first. Another note: Ten years later, I was a soldier in the Ohio Army Guard, flying with an air cavalry unit in a blizzard. I had been released from the storm by the power company, and as soon as I got home, the army activated me to fly for five more days. Remember the bombers' moon during the ice storm? Well, on this one-night flying in a UH-1H helicopter, we could see the shadow of our helicopter on the snow because of a full moon. We were at one thousand feet. I thought back to when we climbed that ninety-foot pole in 1965. Of course, in the helicopter, there was just a little heat, and the floor had some ice on it, but it was safer. A lot safer.

CHAPTER 8

An Interesting Coworker

• • •

VERN M. WAS AN EQUIPMENT operator and had been on the cat crew with Denny, the fellow who got killed on a Saturday night. Vern liked to whittle and draw pictures. He got in the habit of drawing anything that happened in the line department. It was funny. The next day, there would be a picture of whatever on the bulletin board. For quite a few years, the first thing anyone would do is go to the board to see what Vern was up to.

I saw a picture in *Power Linemen* magazine of a fellow who did similar things and found something Vern did for one of our famous foremen in an old company magazine from DP&L that was now discontinued. It was dated 1971.

Notice the following picture. That was Joe R., a foreman who was retiring. Notice the fellow on the pole; it looks just like Joe. The Drew Carey glasses stick out, don't they? Vern had a pair also.

Vern M. putting the finishing touches on his pole carving

Retiring foreman Joe R. with the finished pole carving

My first day as a transmission lineman was in April 1972. At the time, I was a skydiver. This next picture was on the board when I got to work that first day as a transmission lineman. My nickname was Spooks (there's a story about that later). The big bucket was the ninety-foot bucket we had.

Next is another picture of Vern's drawings when I was learning to fly helicopters. When you are on Interstate 70 driving east out of Indiana, there is a big "Welcome to Ohio" arc over the highway. That is the arc in this picture. Notice the rotor blades bent down. Supposedly we had just flown under that thing.

Also, notice all the drag that my big size-15 boots are making. They would probably have slowed down the helicopter by ten knots. When I was an apprentice lineman and three or four linemen were going to climb at once, my foreman would say, "OK, boys, shoe soles and assholes is all I want to see. And, Arnold, just shoe soles for you. Your feet are too big for the other." Everyone would laugh. In the US Air Force, I couldn't get shoes until the seventh week, when we were shipping out. Anyone not having the full issue of uniforms would be held back. I ended up taking a smaller pair of shoes that hurt my feet just to get out of Texas and go to Colorado. Take note, Byron Dunn. (Dunn is the famous retired lineman who is the editor of the fine *Power Lineman* magazine. He is from Colorado.)

I inserted this next picture in to show my feet hanging under that parachute. I was so busy watching the cameraman that I ended up in the fence in the background. This is my picture from 1971.

CHAPTER 9

The Twentieth Jump and a Safety Message for All

• • •

In April of 1971, I was a skydiver at Green County Sport Parachute Club in Xenia, Ohio. We were jumping from Cessna 180s, great airplanes. It was my twentieth jump. We were doing sixty-second free falls. At the jump run, we had to go around again because of a bad spot, so my friend who was flying the airplane continued on climbing. This would give us more time in free fall. As we jumped (four of us), it was not a good jump for me, and I waved everyone off and dumped my first chute, a T-10 Double-L canopy from the US Army. The chute Roman-candled—it didn't open. I thought I had a line twist and tried to unwrap the lines. Not a good idea. By then, I was in slow motion. My hand came in and popped the two-pin reserve chute in front of me. Nothing. In slow motion, I saw my right hand go to the right, come back, and hit the reserve chute, forcing the white chute to bypass the unopened chute and open. As soon as it opened in two seconds (line stretch time), I hit the ground. When I got up, the two-pin ripcord was in my hand. I had ripped it off the bag.

I was in the middle of a muddy field. The chute opened at two hundred feet above the ground. If my buddy hadn't climbed more that day, I would not be writing this story. Note: I still have the lowest jump at that club and am still alive. I fell over a mile that day.

I have thought and thought about how I stayed alive. I must have been spared in order to end up doing something positive. Maybe teaching a lot of new pilots and new linemen to stay alive by always teaching them to listen to "the voice" or their common sense was the reason. There were many more things like that, such as being spared when someone tried to kill me by putting a knife cut in my rubber glove just before I did a very hard job with twelve thousand volts. (That is a story for my next book, maybe titled *Another Peyton Place: Line Department of the '60s.* No promises.) Linemen are always checking their rubber gloves as a hole in them is deadly. I was climbing the pole when I stopped and checked mine that day. That saved my life.

Quite a few times, I was spared without so much as a scratch for thirty-three years. Then, after the power company, I was able to teach flying for a total of twenty-five years. I taught a lot of pilots to hear and listen to the voice, their common sense telling them to do something. I teach every student and all my grandchildren this. It's also a great rule for linemen to follow. If something doesn't seem right, something ain't right. And I use "ain't" on purpose. Do something. Replan, check gas, check around your car, and lock your doors. Lastly, always check your rubber gloves before every climb.

I flew with over twenty-five new military pilots plus many more civilians. I feel lucky to have been able to do this. I have taught over five hundred students in ground school (mostly navy). I show the ripcord a lot, and after telling this story to a class a few years ago, an ensign asked me, "Did the chute open?"

I said, "I am here, aren't I?" We all laughed.

Maybe my mission is to help our country find people who want to join the linemen profession. I hope this book will help America with that problem. The late '60s and early '70s were the times when we started getting better equipment. It was also about the time we realized that our job was getting safer. Before that, we just kept on plugging away, not really thinking about the more dangerous way we were working. Again, thank God for the buckets. You might hear this safety speech again to enforce that kind of thinking.

CHAPTER 10

The Riots on West Third Street

• • •

The time was summer 1968. Ernie A. was back from the army. He went to 'Nam, made staff sergeant in two years, and also got a lot of medals. This was still the time when the temperature was over ninety degrees for two or three days, and transformers would start blowing up. For about ten years in a row, we had that problem, as many people were buying air conditioners and central air units. At first, the transformers couldn't handle the load. So we spent many days adding bigger transformers. Thirty-six-hour days were not uncommon. And it was always very hot.

Later, the company had to upgrade the wire and voltage as they maxed out the capacity of the line with so many transformers. One time after putting wire up, we watched the wire get so hot that it sagged down into the wire below it and blew up. We were still up the pole when that happened. We scrambled to get down and away from the line. It was a little scary. That was when the company had to put in bigger wires to handle the bigger loads and demands that air conditioners presented. That took a few years to do.

So there we were: Ernie A., freshly back from Vietnam; John, a combat BAR man from Korea; and me, a US Air Force vet (not bragging). We were hanging transformers with a bucket truck—1016 again, the oldest bucket that DP&L had. We were up north and received a call to go to the west side of Dayton near the soldiers' home.

It was about 1:00 a.m. or later. To cool down, I rode up in back of the truck, leaning on a bucket. The front cab was small and hot, so it was good for Ernie and John also. So there we were, going to another transformer job. We got down to the center of Dayton and headed west out to Third Street. As we crossed over the river, I saw all kinds of lights and fires and heard a lot of noise, two or three shots. As we got closer, people were breaking house and car windows and throwing bottles; you name it, and it was there. We were driving right in the middle of all this mess. Today, it would look just like Ferguson or Baltimore. It was scary, and all of a sudden, a rock landed behind me. I heard, "Let's get the mothers," and I jumped down into the bed of truck and grabbed my lineman hammer out of my tool belt. I was ready for someone to jump in back, and I'd hit him with it.

All kinds of things were coming in back of the truck, but we kept going. "Go, John! Go, John!" Oh, yes, one more thing: I tried to get in front, but they had the doors locked. Finally, I got in. When we got to a safe spot, we stopped to talk.

This happened at Williams Street and West Third Street. That is where the Wright Brothers Bicycle Shop is. You know, the men who invented the airplane. It's the building where they first came up with the idea of flying. It is still there, and the new area is very nice. When I retired from the power company, I painted the whole Wright Brothers' building for free to help the aviation trails of Dayton. It took me one month. I was by myself most of the month and never had a bit of trouble.

Back to 1968. Ernie, just back from 'Nam, said something like, "Damn, I thought I was out of the war zone." John was going to go on to the job about a mile from where we had the trouble.

I said, "No, I am not going up in the bucket or up a pole and have the lights on me while up there." I heard the voice and listened to it. It was too easy to get shot at. Ernie thought that was a good idea. It ended up that the company pulled all crews off the west side of Dayton that night.

Welcome home, Ernie! Ernie had turned around a bad thing and made it work for him. He came out of boot camp an E-5 and came back from 'Nam a staff sergeant with a lot of decorations. It was then that the guard offered him an officer slot. Later, he helped me enlist in D Troop. No, not the TV show—a real-life air cav unit with real people and real UH-1H helicopters, plus many heroes from Vietnam. I was honored to meet them all and got to fly army helicopters.

On February 11, 2014, I found out that we had passed a state highway patrol officer that night going east to the Dayton jail on Third Street. He was an army pilot friend of mine named Sam, whom I met later. Small world.

CHAPTER 11

The Summer of 1969

• • •

In the summer of '69, I was a first-class lineman at a power company in Dayton, Ohio. At that time, all over the country, many people were upgrading their homes by installing central air conditioning. With this increase in power usage in hot weather, there were a lot of power outages because the lines were not built for every home to have that much electricity usage. Mainly, the transformers would blow. When the weather was over ninety for two or three days in a row, we really got hit with blown transformers. This one night on Rahn Road in Dayton, we had two transformers go bad. We had about twenty homes out. Poles were in backyards on property lines, feeding houses on both sides of the line to two different streets. The electric had gone out just before dark. This is important.

The crew I was on had a driver, a foreman, two first-class linemen, and a younger lineman to help with ground work. We made up a new transformer, rolled it in back on a cart, and got all the equipment in the backyard. The other lineman (Bill L.) and I climbed the pole; I was on top. It was a forty-five-foot pole, so we were up high enough to catch a nice breeze. It was always cooler up these poles. To make a long story short, we got the transformer changed out in good time and were ready to close the switch to energize the transformer. It was now dark.

When we shut the switch, everyone's lights came on. Customers clapped their hands, and everyone was happy again. I looked over to one house, and so did Bill. There, in the upstairs bedroom, was a couple making love. Their lights had been turned on before the outage. It was a good show for a few moments, but both the girl and the guy rolled off the bed and onto the floor, and then we saw an arm come up to turn the lights off.

We ended up working all night—a twenty-four-hour day again—but we had something to talk about for the rest of the night. Just another day at the power company.

CHAPTER 12

Three Good Ones

• • •

Moving Day

Another time in the mid-sixties, Duke was moving again and asked a buddy, Louie M., "Little man, I am m-m-moving tomorrow. Will you help me?" (Remember, Duke stuttered.) Louie had a truck, and knowing it wouldn't take too long, he said OK. Duke didn't have much furniture; he always rented furnished rooms.

The next morning, Louie showed up, and they loaded all of Duke's stuff into the truck (ten minutes' work). Louie asked, "Where are we going, Duke?"

Duke got in, pulled this newspaper out of his pocket, and said, as he was pointing to the paper, "Let's t-t-try this place first, l-l-little man." Of course, Louie had to laugh. By Monday, the whole line department knew about it.

New Carhartts

I was always kind of clean-cut, even when wearing work clothes, and I washed my Carhartts more than a lot of the other fellows. Bib overalls were used to climb poles in; they lasted longer if you didn't wash them too much.

One day, I brought in a new pair of bib overalls, got the dirty pair off the truck, and threw them away. My buddy went over to the trash, pulled out the pair I had thrown away, held them up, and looked at them. He then laid them down, took his pair of bib overalls off, threw them away, and put my old pair on. That was funny, and we all laughed. Another fun thing on the line.

Pigs in the Backseat

On April 3, 1974, a tornado came through Xenia, Ohio. One year later to the day, there was another one that came in just eight miles south of Xenia. It was a small one, and we worked thirty straight hours on the site as power-company linemen. In the morning, at about 4:00 a.m., the Red Cross brought coffee and doughnuts out. The doughnuts had June 6, 1944, stamped on them (D-Day) and were not too fresh. They were not that bad but not too fresh either.

Our boss had a four-door company car. When we had a moment, we opened up the two back doors and tried to get these two pigs that were in this field into the backseat with the doughnuts. "Here, piggy. Here, piggy." It was funny. You had to be there.

Out in a field where everything was gone and was full of mud—it had been raining, and we had been up for almost twenty-four hours—and we do stuff like that. Of course, the boss had to laugh at those pigs in the backseat, but in the end, the joke was on us as we had to clean up back there. Well, I did; Bill got out of that one like he always did.

I wish they had cell phones back then. I'd have a picture of those pigs looking out of the window of that car.

CHAPTER 13

The Day I Set the West Side of Dayton on Fire

• • •

Somewhere around 1968, on a winter afternoon, we were still climbing most poles on the street. We were at Olive Road, about three miles west of downtown Dayton. We were climbing a ninety-foot pole about forty-five feet and were going to add some new equipment. I went up and was ready to open three switches, called cutouts, under a high load of voltage. We had a special tool called a load buster to open switches under load. You hooked in the top, clipped into the ring of the switch, and it would blow the flame out the bottom when you opened the switch. It was a great tool. I opened up the first one OK. The second switch was in the middle of the pole about eight feet above me; I was using long sticks with the load buster on the top. As I opened the second switch, the top of load buster came off the top part of switch.

Of course, the flame followed and came down the stick about four feet. The stick went over to the left side of me, and I let it fall as it was time to hit the dirt. The flame was twelve thousand volts. I put both arms above my head after a piece of the switch knocked off my hard hat. I saw the flames blow from the middle of the pole to the far west side of the pole, over four feet out—all flame. I was afraid it was going to go up higher as there were sixty-nine thousand volts eight feet above this equipment. Finally, the flame reached the outside of the cross arms and went phase to phase on two 7,200-volt wires. The damn thing tripped out the feeder, and the major part of the flames was out. When it was burning, the wind blew the fire from the middle wire to the far outside wire, over four feet, just eight feet above me. I was OK except for a cut on my head from broken switches knocking off my hard hat. I looked down at the snow on the ground, and some of the switches were still burning while lying in the snow.

The cause? The load breaker malfunctioned just as I opened it. We cleared up the burned-up switches as fast as we could, since we had many big companies like GMC that were out of electricity. It took over an hour, and when we got everything back on, we made up material for the next day's work. The next day, we completed the work. It was just another day on the line that only a few know about.

The linemen of the '60s really did have a harder job to do than I realized at the time. The bucket trucks have made the job a lot safer, and I am glad. Another thing that I am glad about is that linemen make better money than they did when this happened. I was still just making just over four dollars an hour. Twenty-five years later, my highest pay was still low at $13.50 an hour. Now they make thirty dollars an hour or more.

The next day, we were back on the same job, and we changed to a larger capacitor bank and all new switches and arrestors. These are items that go with each capacitor. All of these types of poles are now done with bucket trucks for safety. All of the junk I had was above me that day as I was on the pole. In the bucket, you can swing away from bad things.

It was about that time when I learned that the average life span of a power lineman at the time was seven years. That's right where

I was: 1961 to 1968. That old rule is out now as the bucket trucks have made the job so much better. I stayed for twenty-five more years. Then I got to teach flying for twenty-five years.

Sometimes when a student had me upside down at five thousand feet, I missed being up a pole.

CHAPTER 14

My Buddy Jim (Gym)

• • •

I STARTED TEACHING SCHOOL SAFETY programs for DP&L, the local power company, in 1987. I was working from four to midnight and going to schools in the mornings from eight to noon. All in all, in two years in Fairborn, I taught over two thousand kids about wires falling down and draping over a car or things like that. They got a kick out of the shotgun tool we had; they all liked the idea that it also helped to keep doggies away from our clanking tools when we walked by putting out a high-frequency noise that dogs don't like.

At some point, I got the idea of making a dummy. In Fairborn, Ohio, where I was working at the time, there's a famous toy store called Foy's. That's where I went to buy a (bad-looking) rubber mask. I had a picture of the dummy and me with the truck; Mr. Foy, the toy-store owner, couldn't find it, so I am making a replacement Gym. He will be wearing the new DP&L blues. Anyway, back to the story.

I stuck a Styrofoam head on a short broom handle, put a company baseball cap on it and an old rain jacket with a hood, put the hood over the baseball cap, and I had half of a dummy. The kids named him Jim, but I put "Gym" on his nametag, and all the kids got a kick out of that.

At night, I'd see a lot of these kids as I was everywhere with the trouble calls, and they all would ask, "Where's Gym?" So after a few questions like that, I rigged up a way to get old Gym in the truck with me. I could even reach over before my shoulder went bad and turn his head if I was in traffic at a red light so that people in a car beside me could see how handsome he was. Sometimes a good-looking girl would stop beside Gym, and I would reach over and turn his head so he was looking out. Some girls would tailgate the car in front of them at a red light just to pull up more to get away from Gym. Poor Gym. That always hurt his feelings.

The daytime linemen used the truck while I was off and would sometimes hide Gym. One time I couldn't find him and had to go on an outage trouble call. When I got in the bucket, there he was.

Here are a few things that happened with Gym. One night, I pulled into a UDF store to get coffee at around three in the morning. I pulled the air-brake lever. You know that sound? It hisses, and when it makes the brake set, it is noisy. Right at that time, there was a guy just going into the store right in front of the truck. He had a gun in his hand. Just as air brakes held with the loud sound, he turned around and looked straight at Gym, who gave him a bad stare. He took off running. I hadn't seen the gun until the brakes set. Of course, we called the sheriff. Plus

I got free coffee there for quite a while.

Another time, two guys in a car stopped where I was refusing a streetlight pole near Interstate-75. I think they were going to rob me. I yelled, "Gym, get out here. We have a problem." They must have seen his head because they took off.

Going to fire calls, I always drove fast with yellow lights flashing; this was usually sometime around 3:00 or 4:00 a.m., when there was no traffic at all. Once, I passed this drunk guy who was on the sidewalk, and he was petrified of Gym. He held his arms over his face and screamed as I went by. I had to laugh.

Here's the last one. On a Sunday night in the west side of Dayton, I had a branch-line fuse blown. A branch line is a 7,200-volt wire feeding another section of wire. Just like with fuses in your house, if the circuit gets in trouble, it will blow the fuse. So I was working to refuse this switch. We had a forty-foot stick, so I could replace a fuse while staying on the ground. Most linemen could really do a good job with these telescope sticks to save themselves a climb or getting in the bucket.

It was a very hot night, and with all the power out, everyone was out of their homes to get cooler. It was very dark, but there was a full moon, which helped a lot. As I was using the forty-foot stick to refuse the switch, there were about thirty people around me, and I had to ask them to move back some in case I dropped something. This woman asked me, "You got a dummy in your truck?"

I said, "No, I don't, lady. He's smarter than I am. He is in there sitting, and I am out here working." Everyone laughed.

When I closed the fuse, it blew like a 105 cannon. I mean it was loud with a little fire. I looked around, and everybody was gone.

I mean everyone. I never saw them again. I found the trouble: a wire was down two blocks away. Another crew showed up, and we repaired it pretty quickly.

Finally, Gym had to stop riding with me. He was making more overtime than all the other linemen. Also, women were calling into the company wanting to meet such a handsome fellow, and my buddy Mike H. in Fairborn didn't like that. (These stories are all true except that last part.)

After I retired, I rebuilt Gym to sit in our front yard for about a week before Halloween beggars' night. The kids got used to him being there. On beggars' night, I dressed up in the same outfit, and when the kids came to the front door, I would move and have fun with them.

This was Gym after I retired, playing with the Halloween kids. It's not a good picture, but look at the feet.

CHAPTER 15

Nicknames in the Line Department

• • •

For some reason, there were a lot of nicknames in the line department. We had a Froggie, a man with big eyes; a Biggie, a man who was as kind as he was big; and an R. J. Dunn, a fellow who had the name Bill Dunn.) My nickname was Spooks because when I first started climbing, my mother gave me a blue face mask with white eyeholes and red around the mouth hole. I only wore it one time, and, of course, I couldn't see well enough to climb with it on, but all of a sudden everyone started calling me Spooks. My mother hated that name, but it didn't bother me. We can thank Paul C. and Gene B. for that nickname. Their nicknames were Brutus and Cashes.

As transmission linemen, we traveled a lot. We worked south of Dayton at the Ohio River site of a power station, Stewart Station. We were there quite a lot from 1972 to 1973. I had a small coffee pot in my room and somehow got in the habit of waking everyone up with a half cup of coffee.

A big reason was that the foreman, KB, would wake me up wanting coffee. So after we had a full cup, I'd share the rest with the crew as I went from room to room, waking them up. Well, when I started doing that, the crew started calling me Mom. It stuck, so now I had two nicknames, Mom and Spooks. When we were back at home base, I was both. But, later, as a leader, I started getting called Mom more, mostly by the younger linemen.

One time, the crew had to stay over the weekend at the Ohio River so we could get a clearance on the line on Saturday to work it de-energized. We kind of partied hard on the Friday night. I remember putting on a shawl that had been hanging on a clothes hook for two weeks and had not been claimed. Going back to the bar, I saw that some guy had taken my seat. I told him that he was sitting in Mom's seat, and he got up and said, "I am sorry."

I said that was OK and sat down.

He said, "I thought you said 'Mom.'"

I said, "I am Mom."

He started to get the red ass —a common saying in that area, but KB came up and said, "Leave my mom alone." It was over, except we stayed too long that night and ended up having a double order of coffee in the morning. Boy, that was a long day.

A note about the shawl: It was in my room the next day, so I took it home and gave it to my mother. She asked me where I got it, so I told her, "I picked it up at the Ohio River." She never said another word. She had it until she passed away.

One friend who was also a transmission lineman started sending me Mother's Day cards on Mother's Day after we both retired. I got one yesterday. Thanks, George. Love, Mom.

CHAPTER 16

Six Calls in Thirty-Three Years All by Interstate 70 within Five Miles

• • •

FOR SOME REASON, I HAVE a great memory for the dates when things happened in my life and places I worked as a lineman at DP&L for thirty-three years. I have no idea how I can do that, but I can. Now that I am not flying, my wife and I are going to Columbus, Ohio, one day a week to try our luck at a newer casino. It gets us out, and we have fun. Going east on I-70 from Route 235 to the Route 4 exit, which is about five miles, there are six different sites where I did something for DP&L. Here is some information on each one over thirty-three years.

1.
The first was as a laborer at DP&L. We were required to help on the pole delivery truck in the spring of 1961. Jim F., another soon-to-be lineman; a fellow named Ned, the pole truck driver; and I delivered poles at a spot just east of I-675 and eastbound I-70, just past the big tower line on the right. They were ninety-foot poles, so they are still there in a 138,000-volt line on double poles with an H construction. Later, in the '80s, I changed one out with the ninety-foot bucket and Mike V.

2.

At the same location, in January 1975, at the 345,000-volt line, I alone climbed a 225-foot tower to the very top (crow's nest) and made a connection of copper to the static line and clamped it to the tower for proper grounding. I carried up the connection press, so I didn't need a hand line going by the 345,000-volt line that was energized. I had my grounded boots on up in all that static electricity. Boy, were my legs tired after that climb; it was muddy, and that added to the weight. I airmailed that press back down (threw it) 225 feet, and it stuck in the mud pretty well. Of course, the foreman said something. I told him to bring up a long rope, and I would let it down again. All that wet rope around the 345,000-volt wire was something I needed. Duh! Normally, I wouldn't do that, but carrying that press down that tower was not a good idea.

3.

Just east of that location is a three-pole, double dead-end setup that we worked on in 1973, just after I joined the guard. When we were transferring the wire from one pole to a new pole, the heavy chain-wire pullers slid down, and the wire started going down over the highway (I-70). The foreman went nuts and started yelling, "Get the tuggers! Get the tuggers! It took a minute to get the bucket to the wire, but we stopped the release of the wire from getting lower and lower. In the end, it was kind of funny, or so my friend George and I thought. The foreman was pointing when he was yelling. I was doggin' it. Another new saying which means goofing off.

4.

In the fall of 1963, at the end of my two-year stint as an apprentice lineman, there was another job with the same wire across the highway and going north. The company had every line truck in Dayton and other areas to add insulators to each wire and beef up the voltage from sixty-nine thousand to 138,000 volts. We did this by adding, I believe, three insulators to each wire. It was an H-fixture setup. We two linemen, Jim W. and I, climbed the ninety-foot pole and did the east side first. We used an eight-foot wooden ladder called a hot ladder, with big hooks on it. We put it on the timber, stepped off the pole to the ladder, and rebelted our safety through the ladder. We worked the wire, added insulators, got off, moved the ladder to the middle wire, and redid everything. We got off again, moved the ladder to the far side, and walked over the middle on the beam. The ladder and all the rope went with us. We worked the last wire and dropped everything down to the ground on a hand line. Then we climbed down the second pole of the H fixture, up one and down the other. We did eight poles that day.

About two weeks later, one of the wires came down on that line up north, and a farmer was killed. The company made up the same crews that had worked the line before, and we had to climb every pole again and check each insulator to make sure that it was pinned right. There were over a hundred poles on that line, so there were many men on that job. We didn't find any insulators on our poles that were not proper and installed right. We were glad about that.

5.

In 1990, as a line trouble man, I worked on a big power section out of electricity just east of 235 on I-70 at Mad River. I got there, and a serviceman was already there. We walked into the woods; it was just before dark but still light enough to see well. We found a big tree down on all wires between two poles. All of the wires were OK, but it was all together. I had a forty-foot stick with me that we could use to change fuses on the ground and not have to climb a pole. I put that stick on the wire and started moving the wire up and down. Now this surprised me. I was just trying to get the tree to slide off the wire. The wire was rocking back and forth; the tree was about six feet in diameter, a big one. The damn thing completely rolled down all of that wire and fell to the ground, and everything flipped back up to the normal position. Talk about being lucky rather than smart. We called the station supervisor on the radio and had the whole line back on in about fifteen minutes. No one in supervision said a damn thing about what a great job we did or even what happened. That's what happened in those years. There were no experienced leaders in office—just my buddy George R., and he was about ready to leave early and forever. I still miss my buddy. In a few days, it will be Mother's Day, and I'll probably get a card in the mail from him.

That timespan stretched over thirty years, and each job was just within a few miles of each other. I didn't even talk about one where I did the whole pole and always say, "That's my work" to all my friends as we go by. I figure that they are tired of me saying that, so I didn't report on that site. But it was a great pole (haha!). Claude, take note. Claude is my flying buddy.

6.

Around February in 1989, I was on a single-line trouble crew doing everything on the four-to-midnight shift. An Ice storm had started, and we had reports of the wire blowing up just east of 235 and right beside I-70 eastbound. I got there after taking a while, for the roads were very bad. I pulled over to see what was blowing up and had turned all the lights off. I was completely off the road. All of a sudden, the sky lit up with about fifteen blue lights from the top wires to the ground in about three different spans of wire. It happened about three to four times and then quit.

Now think about this: small blue explosions that many times? Well, this was an ice storm, and when the ice covers everything, it creates an electrical path, as the ice has molecules or iron in it. It's not pure—kind of like an army aviator or a lineman! Well, about two or three days before, it was a very windy day, and I had seen kids flying kites in that location, and there were a lot of kites in that wire at the top. There was a lot of string blowing around in this wire. So each time the blue lights blew up, it was going down that string to the ground and—poof—no more. I called it in and was sent somewhere else for another twenty-four hours. Actually, it was kind of pretty. That type of call was all over the whole area that night. The very next weekend, I went by that location, and there were many kids there again flying kites. It was another windy day in Ohio.

CHAPTER 17

Meet Me on Southbound Interstate 75

• • •

In the late '60s, a three-man leader crew was supposed to drive south on southbound I-75 until they saw the crew they were supposed to help. The foreman, Joe R., had told the leader of the three-man crew, Taylor C. or TC, to just keep going south until they saw each other. I think he had forgotten the name of where he was going and didn't have the map with him. Froggie (Roger K.) and John J. were the two linemen on the crew. Froggie had been partying the night before, so he fell asleep. When he woke up, they were still driving south, passing Crosley Field. They were in Cincinnati, Ohio, fifty miles south of Dayton, right by the Cincinnati Reds baseball field at the Ohio River. Froggie said, "That looks like Crosley Field."

John said, "It is." They drove to the Ohio River and turned around. Everyone laughed about that when they finally joined up. It was a different time.

TC had been a marine in Korea and had seen a lot of bad things. Once he had mentioned to me about being in a bad fight on Christmas Day. He choked up and stopped talking. He was another good man who could have used some help.

Later, we found out that Joe's crew was delayed in getting on the highway, and TC passed where they were supposed to meet. No one was there, so he kept going just like he was told to do. Those were the good, old days. Joe R. was the fellow who retired with the wooden carving of him up a pole.

CHAPTER 18

First Aerial Electric Lineman of DP&L

• • •

In 1979, Dayton, Ohio, had a very bad blizzard. All electric linemen for the company worked for many hours. The biggest problem was getting to where the outages were located around the Dayton area. The snow had most of the highways blocked, but the fields out in the country were mostly clear as all the snow was on the road. Some of the roads had snow ten feet high with cars under it. The next week, with the Ohio Army National Guard, we would land in the fields in helicopters, climb up on top of the snow, and work our way from car to car, making sure that no one was in them.

At one point, the company rented a Bell 47 helicopter and a pilot. There was an outage west of Dayton near Eaton, Ohio, and it was determined to be either a branch line or a transformer fuse that was out; either way, it was a pretty big section outage. I remember that you could not drive to the fuse pole because of the snow on the ground. So with a forty-foot fiberglass stick and fuses in hand, Froggie was flown over the hilltops of snow and landed near the pole. Soon, he had everything back on, and he became DP&L's first aerial lineman.

Once released from the storm efforts, two linemen from DP&L returned home to find out that they had been activated by the Ohio Army National Guard to fly support missions for the State of Ohio. Those two linemen were me and Steve P. It was quite a winter that year, and it was also scary, landing by the roads in a field that had no snow in it. All the snow had blown over the roads. I would get out of the helicopter, walk over to the road, climb up the snow probably about eight feet high, and go from car to car to chip away the ice to make sure that the car was empty. I never saved anyone, but up north, a guard unit found a trucker who had been trapped for three days. We were released after five days.

That was the time of the year when we were all ready for spring and a nice, easy thunderstorm to work instead of all that ice. Ice storms are the worst and also the most dangerous, even with the buckets. It's safer climbing with the buckets, but the ice makes everything so damn heavy. Many times, we would put up a lot of wire above the trees that were sagging down below our new wire because of the ice on them. When the ice melted, the trees would shoot up right back into our wire and burn it down again or at least blow the fuses. That is why, a lot of times, the tree trimmers cut down almost everything. Trees and wires do not get along.

There is a small town on Route 48 just north of Englewood, Ohio, that has some very nice streetlights on both sides of the road. But in line with the power poles, they have planted some very nice trees right underneath the wire. In about five more years, they will start having trouble with that setup. It's just a matter of time. That is the same thing that happens in housing plats. The planners plant all of these trees, and guess what? In about ten years, they start having problems.

CHAPTER 19

Your Side

• • •

In the early '60s, Duke, Tiny W., Jim S., and Poopsie (the driver) were working storm trouble, putting wires up after a storm. Well, Jim had to leave as he was getting married that night. I do not know what time the wedding was. Let's say 7:00 p.m. So he left, and the rest went on working. Well, at about seven that night, the storm was almost over, and the crew had been working close to where Jim was getting married. So they all decided to go to the wedding instead of getting something to eat. Picture these guys—a little bit grimy, in Carhartts clothes that workers wear, and a little tired and dirty—going into the church. Well, they all sat on the bride's side of the church!

Jim and his bride got married, and everyone filed out of the church, shaking hands and saying nice things to the bride and groom. Duke got up to the bride and flat-out planted one big kiss on her in front of everyone while saying, "Good luck, sweetheart."

After they left, the bride asked Jim, "Who was that?"

Jim said, "Don't ask me. They sat on your side of the church!"

As a side note, Poopsie got drafted during the Korean War. He was told to report on the other side of the room. After he was sworn in, he asked why he was on the other side of the room. He was told because he was going to Paris Island. He was a marine—a good one too.

Jim, the bridegroom, was killed in 1962. He was the helper for LR, who I worked with my first week as a second-class lineman. That was a very sad time.

CHAPTER 20

Call-Ins

• • •

DP&L linemen were always on call. When the phone rang at 3:00 a.m., and it was snowing or storming outside, you knew who it was. We kept track of everyone's overtime hours on an overtime list. The person with the lowest overtime hours was called first. If he turned downed the overtime, the next guy would be called, and whoever worked, say, four hours double time, they both would be charged with eight hours on the list. This was done to try to keep everybody equal on OT. The first guy had a chance to work it and didn't.

As I was single and spent time with my boys on weekends, I was always being charged for overtime, as I couldn't leave my boys alone. One day in the '70s, I got a call from Dutch A., a very great supervisor at DP&L. Everyone liked this man. When he said, "Is Bill there?"

I knew who it was, and I said no.

He said, "You sure sound like Bill."

I said, "This is Forrest, Bill's twin brother. Bill is at the airport, and I am watching his boys."

He said, "OK, tell Bill that DP&L called."

I said, "OK, Dutch" and hung up quickly before he could say anything. He never, ever said a word to me about it, and I didn't get charged for the OT.

Dutch was a great fellow, and I just told him this story about two weeks before he died. We both laughed. He was another WWII army communication sergeant. As he was already a lineman when he got drafted in WWII, right after basic training, he was made a corporal and taught people how to climb poles. Then he went off to Europe.

Here's another OT story. This one lineman was held on a job after the normal eight-hour day. After eight hours, we went on time-and-a-half pay up to midnight. Then it was all double time. All call-ins were double time. So this fellow was working, and there was more trouble, so they needed more linemen. The call-in supervisor didn't know that this one fellow was already working and called his number at home. When his wife answered the phone and the supervisor asked, "This is DP&L. Is your husband there?" she said, "He'd better be there already." It was sorted out.

Another time, this one crew was held over after a normal day and finished the work at about 8:00 p.m. They were supposed to get a meal with pay after that long. This crew decided to put down another hour on time and go on home. It always took an hour to eat, so that was fair. But one fellow got called back in as soon as he got home at quarter after eight. So now he was getting time and a half and double time for the same forty-five minutes. He worked with me that night, and I asked him, "What did you do, pass yourself on the way to work?" Of course, we all laughed. No one ever questioned it. His pay sheet went in OK, and he received that amount.

CHAPTER 21

Orange Balls

• • •

I SUBMITTED THIS STORY TO Byron Edgington for his book *Sky Writings* since it involves safety in aviation. This is a lineman story, but since I am also a helicopter pilot, I sent it to Byron because I was sure that some of his readers flew over this area. We did in the air shows in the '80s with the guard. Some of Byron's readers have probably flown linemen to install these orange balls from helicopters.

In the late winter/early spring of 1979, three of us with the ninety-foot bucket truck were sent north of the Dayton airport to install three orange plastic balls to the top static wire of the 345,000-volt tower line. These orange balls are markers on the wire for pilots to see and avoid hitting the power wires—or at least we hope. But they were mostly for helicopters. This 345 line is about three or four miles north of the approach end of runway 18 at DAY, which is the identifier for the Dayton International Airport. I could tell you more, but I'd have to charge you. That's flight-instructor humor. Sorry.

We arrived at the job on North Dixie Drive underneath the 345,000-volt wires. We would have to get the bucket down a muddy field about five hundred yards to get to the job area. We had a bulldozer pull the ninety-foot bucket through that muddy field. It took quite a while. We finally got the bucket set up to do the first marker. We were putting up three of them.

We got our clearance on the 345,000-volt line, which means they took the line out of service, or de-energized it. We still had to feel out the line to make sure it was not hot. It was not, so we grounded all the wires together, which takes a little time. It was at this time that we decided we had a little problem.

The problem was that we couldn't get the bucket high enough to reach the top wire that we needed to get to in order to do the job. What should we do? It doesn't take too long to put these markers on the wires, so we decided to put a small, ten-foot underground ladder in the bucket, lay it against the top wire, climb up, and belt off to the wire to secure us as we installed the markers. The job area was about eight feet above the bucket. So I did the first one, George R. did the middle one, and then I did the third one.

Here's the picture. We had to pull the bucket in with the bulldozer, set it up in the mud, go up all the way (ninety feet), put the ladder in the bucket, lay it against the top wire, climb up eight feet, and belt off on the wire through the ladder rungs to be secure. We did it. And we got the job completed in one day. Even though this was not a proper procedure we got the job done.

This site is about one mile south of State Route 571 and I-75 on the west side of I-75. The three orange balls are still there. I almost always salute them every time I go by. I always think of my friend George, who was with me that day (again). We were just trying to do the job, and we did. It was a different place and a different time. Now, they would use a helicopter with a fellow sitting on the skid to do the work.

One more note: George is my buddy who sends me Mother's Day cards every year. See what I am talking about with the lifelong friends you will make? Plus linemen are finally making the money they deserve now—not then.

CHAPTER 22

Early Storms for Me and More Information

• • •

My first ice storm was in 1962 as an apprentice making $2.35 an hour. Yes, that is what I was making when I started climbing poles. When you add it up, double time was still less than five dollars an hour. Plus, there were only two bucket trucks, and they were not for the apprentices. But you know what? We were in hog heaven and thought that we were getting rich. Time and a half was a whopping $3.53. But the first time I saw people's reactions to us after they had been out of heat and lights for a week, the kindness shown to us was wonderful. All of the young linemen thought that was great.

Sometimes when we first got to an outage job, people might have been pissed, but after seeing what we had to go through to get them back on, they changed their tune. Remember, we were climbing everything, it took longer, and it wasn't as safe or comfortable as the buckets. It was colder, too, and climbing an icy pole took more guts than we were used to using. We all became cockier in that first storm. It was because we found out that not everyone could do everything that we were learning to do. It was a little bit like the airborne attitude—actually, a lot like it. There were a lot of ex-airborne men in the linemen force at that time, mostly the older fellows from WWII, but the apprentices had some also. You could tell which apprentices had been in the military because of the clothes they wore. The company didn't have uniforms for linemen yet.

I have to add a little more information about how much safer line work is now. In 1962, we worked almost thirty hours that first stretch. Then we were allowed to get five hours off for sleeping, cleaning up, and eating. Don't forget the drives home and back. So there wasn't much time for rest. During my first big thunderstorm in 1963, when I was still an apprentice, we worked thirty-five hours the first shift. I was bitten by a dog while I was connecting wire together on the ground. The reason? Dogs don't like the sounds our tools make with the clanging and high-frequency noise, and we were in their yard. That dog put teeth marks in my climbing tools. I had to go to the doctor during my five-hour rest time. So I really didn't get even four hours off, and I still went back out. I needed the money. Little Scott was on the way.

That would not happen today. Now the companies try to do sixteen on and eight off. It might be a little longer to maybe pick up a big section of customers but not much more. You will also get a clothing allowance, which really helps the budget at home. It also identifies all workers wearing the same type of clothes, which is another safety feature. Plus, they reflect green and are easier to see in the dark. People are used to what the power-company employees look like. In 1962 and 1963, we had to keep telling people who we were. Sometimes we had guns pointed at us. Making everyone look the same and wear the green shirts and larger white hard hats makes it a lot safer in the backyards, which are really dark when the power is out. Now, people's attitudes will be the same, but you will be making ten times more money and have much better equipment to work with.

Many of us in the 1962 ice storm didn't have rubber boots. Now the companies issue hot boots to go over the climbing boots. They also have thin rubber boots you can get when it is really bad. Remember seeing young mothers putting little plastic bags on their children's feet to get the shoes to slide into the boots easier? Well, that is a damn good idea, except I had to use big trash bags for my size-fifteen boots. One time in a restaurant, I had on white gym shoes, and this woman said to me, "Darn, what size are your feet?"

I said, "Fifteen, miss, and if you think these are big, you should see my jump boots." People around laughed. I said to my wife, "Karen, get their address. We'll cut off their electric." The laughing stopped. I was just kidding. But it has come up a few times since I retired. People say, "You don't work there anymore." My answer is always, "I know, but I haven't forgotten what I learned." And if it is my buddy Claude, a "slabhead" is thrown in.

One time, after getting off at midnight and still wearing work clothes that had the power company's name on my shirt, I stopped at Speedway to buy some drinks. The lady asked, "Are you the cutoff man?"

I said, "No, ma'am, I am the turn-on man." She liked that, and later we became friends.

Now you can see the kind of humor we all picked up. Well, some of us. One of my quotes is, "Linemen are like paratroopers on Saturday night—but every day." I am telling you all that there were times when we didn't want to miss a day of work because we thought we might miss something. As I've said before, thirty-three years was a blast. It still is, so good luck with whatever job you will be going for in this field.

If you do not want to go for the climbing part of line work, there are still highly paid equipment operators and truck drivers who will be grunts on the job. Grunts are the people who tie things on the hand line while other workers are up a pole. Operators will set the poles and operate almost everything: backhoes, pole trucks, you name it. Remember, linemen can't do anything without the support of these fine people. The bucket trucks have helped the ground workers also. They can go down and up so easily in those trucks that, many times, it can save the grunts the work of putting things on the hand line. The climbers can go down and get it.

So, again, readers, becoming a power lineman will not only help you, but it will also help our country. Good luck, and God bless America.

CHAPTER 23

The Newer Bosses

• • •

SOME OF THE NEW BOSSES came from service centers and had never been around linemen. Even though they meant well, they still didn't understand some of the things that went on. For example, on my fifty-fourth birthday, my helper and I received a call about a large outage north of Dayton and just west of Dayton International Airport. There was a wire down on a van and a fellow trapped in it. We arrived at the job site, and, yes, there was a fellow in a van with primary wire on top of it. I told him to stay in the van and located the fuse feeding the wire. The fuse was blown, and the wire was dead. I flagged the switch pole with a red flag and then kept the fuse holder so that no one would refuse it while we were working on the line. Also, we were only a block away from where the van was. The fellow had hit a forty-five-foot pole and had broken it, causing a large outage.

When we got back to the broken pole, the guy was still in his van, drinking coffee or whatever he had in his thermos. I told him he could come out and talk to the police. To make a long story shorter, my helper and I (sorry, I don't remember his name) took two sets of hand metal blocks and put up the wire temp and had connections all ready that the crew coming out to reset a new pole could use and save a lot of time. We had everyone, except two houses, picked up in less than an hour with a temporary connection.

The crew arrived, and we left. My old roommate, Dave P., was with them. God bless him. He passed away two years ago. He was a very good lineman and soldier. He had been in the Big Red One, a famous unit in the army in Vietnam, and he had seen a lot.

The very next night after that call, I had a blown fuse. I arrived and refused a fifteen-amp fuse in about five minutes. The lady called in and praised me for doing a great job. The next night, the boss handed me a nice letter to put in my file about doing a grand job on the inside fuse job. It was nice but out of place.

No one ever said a word about getting three to four hundred people back on in a short amount of time the day before, which was a great piece of line work. An ex-lineman would have mentioned it. That brings up another positive piece of information: the new bosses are all ex-linemen or have ex-linemen under them, keeping them informed. It is a much better working environment and a safer operation for all. I am glad that was changed.

One more thing about the present day: I think there are many more positive stories about power linemen than there were back in the '60s and '70s. I am glad. I always thought that we deserved more "good job" comments than we received. A little thank-you goes a long way.

CHAPTER 24

The Mickey Mouse and Glass Insulators

• • •

In the 1920s and '30s, this was a very popular insulator. It was used on cross-arm construction of poles for the electric wires. Mainly, it was the insulator that the neutral wire was tied upon. This was first called a "donkey ears insulator." Later, after Disney made the Mickey Mouse movies, the name was changed to a "Mickey Mouse insulator."

In the 1970s, when rebuilding those 1920–1930 circuits, we were around these a lot. We were getting five dollars each for them from people who wanted to buy them. They are one of the better-looking insulators that I have ever been around.

My friend Dave P., who was my roommate for quite a few years, brought home a glass insulator one night and gave it to me. He had been working in my mother's backyard. They had changed out the pole and replaced a transformer to increase the primary voltage to 7,200 volts. The glass insulator was for 2,400 volts and had to be replaced. Thirty years before, I saw linemen climb that same pole and install the insulator.

Insulator that was on a thirty-five-foot pole
behind Bill Arnold's boyhood home

CHAPTER 25

Wires Shot Down

• • •

WORKING THE TROUBLE EMERGENCY SHIFT at the power company on New Year's Eve of 1992 was a different kind of trouble shift. Every electric trouble call we had during that shift was about wires having been shot down. I was working with a helper named Randy, a great fellow. I just saw this man about two years ago. He still looked great and was on a bucket truck that had some really nice tools to do a safer job.

Back to New Year's Eve 1992, the first call was about one wire of a three-wire service going into a house crossing a street. The call was a part-out call. As the street was heavily traveled, Inspector Clouseau (me, haha!) devised that this wire was shot down from a moving car. We put the wire up and had them back on in less than an hour; the traffic slowed us down a little. I was right about it being shot down. You could tell that the cut was rounded, not even.

Later that night, we got the second call from a house right across the street from the first call. This fellow was frantic about people being in his backyard and shooting. He was drunk but was a nice old fellow, so we decided that since we weren't too busy, and it wasn't a bad night for New Year's Eve weather-wise, we would put up his two-wire service going to his garage. The main reason was that he had a refrigerator with a lot of food in it in the garage.

Just as we were setting up the ladder, the fellow started screaming, "My chitlins are burning" and ran into the house. In about five minutes, my helper, Randy, had the wire back up and on.

Putting away the ladder and picking up any wire cut off on the ground, we noticed about twenty twenty-five-caliber casings on his back porch. Again, Inspector Clouseau was back on the scene. There had been no one else in the backyard like the man had said. He had shot his own wire down at midnight. We turned down the offer of having chitlins. It was time for a coffee break and to stand by for another call. But before we left the chitlin man, we figured out that he had also shot down the first wire in the front of his house, and it was not done by a drive-by car.

The next call came just as the football games were starting, probably at around 1:00 p.m. We found primary down that was feeding about fifteen houses. (Primary is the 7,200-volt wire feeding transformers on the top of poles, for people who don't know.) We had to open up the main fuse to de-energize the wire to fix it. Guess what? We turned off the football games in about fifteen houses while doing this. All of a sudden, we had some irate customers out on the porches, all yelling. I said, "Folks, due to one of your neighbors shooting at midnight, and you probably had a blast when the wire burned down, seeing the flash, your game will have a short delay while we repair your wire and make it safe." That call had hot wire hanging down in the backyard, where someone could have gotten into it. They cooled down after my speech, and the games were back on in a short time.

If I remember right, we were working sixteen-hour shifts. After that, it was time to go visit my mother for our last New Year's Day together.

CHAPTER 26

Quotes from Brothers

• • •

HERE ARE SOME QUOTES FROM my linemen brothers. Some are gone, but they are always with us in thoughts. Some things that happened while working were not normal—many things, actually.

Jack R. said, "It ain't supposed to taste good; it's supposed to make you feel good." After putting up wire in an ice storm on a very cold day, we were standing around the truck, putting our tools away and getting our reward: a taste of Canadian Mist. It seems like most line trucks had similar rewards on them. In the summertime, for some crews, it wasn't uncommon to have two water coolers aboard the trucks—one for water and one for beer. Not all trucks had them, but a few did around 1970–80. The company cracked down on that type of habit in the 1980s. They also really cracked down on drugs. You can't use that junk and do line work. You need a clear head to do the work. There were a lot of problems in the '70s.

Jim L. said, "Look at her, boys. She's in her brier and panties." This was from a hot-crew foreman on a hot, stormy night in Dayton when a woman came out on her porch in her underwear to watch us work. Jim was another WWII hero from the airborne division.

Dave N. said, "Shoe soles and assholes, boys. And, Arnold, just shoe soles." Our apprentice-crew foreman said this quote when having the whole crew, usually four linemen, climb all at once. What he meant by that was I had size-fifteen boots, and he couldn't see my ass. Dave was another WWII hero, a communications sergeant in the Pacific.

An underground foreman nicknamed Stoop called in one day: "Dispatch, you'd better call the police in West Carrolton, Ohio. Some car just hit me in the ass." That was over the company radio. This fellow had received a lot of medals from the Battle of the Bulge in 1945. He told me one day, "I am still fighting the battle of the bulge." He was a fine man.

Ed C., a truck driver and friend of mine, had hit a man in a barroom fight and missed the WWII thing as he had been in prison during the war years. He had been in a couple of boxing matches and was deemed a professional fighter. The man died, and that is why he was behind bars for so long. We were on a crew together, and we didn't get along with the foreman, so Ed would ask me how to spell the man's name, and I would spell out "slabhead." Then I'd say, "Ed, how do you spell 'slabhead'?" Then Ed would spell the man's name, and we would both laugh.

I used to be up a pole and ask, "Mr. Ed, have you ever been drunk enough to kiss a woman's belly button?"

Mr. Ed would yell back up, "I have been drunker than that." Then we would both laugh. Sometimes other people would too. Ed was a good friend; he died in 1971. Being in prison probably saved his life. He told me he would have joined the marines in the war. Ed would have been a great marine.

A little boy in a car beside a three-man crew, where all three men sat in front at a red light, said, "Look, Mom, he's drinking a glove." Sometimes the boys would put a beer inside one of their leather gloves to hide it as they drank it. Later, they had stickers to put around the can that looked like Coke labels.

Dick B.'s famous quote was, "A few kind words and a smile." Dick had a silver tongue. Sometimes he would make out well, call home, and say, "I am working late." Then he'd stay out late. Sometimes the next week he'd get extra money from credit union to make up for the overtime pay. He is still a fun guy. Sorry, Rabbit, that I told on you.

Duke C. took over our apprentice crew for a couple of days when our foreman was off. Duke was still a first-class lineman then. He was upgraded to temporary foreman for an extra twenty cents an hour. He stopped in the morning for coffee; this didn't happen every day unless the foreman wanted some. When we sat down at the table, Duke made the announcement (and, remember, he stuttered), "Little m-men, to make up for stopping for coffee, we are going t-t-to lunch early. And t-t-to m-make up for going to lunch early, we are c-coming back late for work. And for making up for coming back late, w-we are leaving early to g-go in." That is what we did that day, which was one of the better days for me as an apprentice lineman. Sometimes it was like being back in boot camp. Actually, boot camp was a breeze compared to what the apprentice linemen put up with in the '60s.

Gary, a climbing serviceman, was working with Mac, the father of a great lineman, Gary M. Mac was at the house with a lady customer, and Gary—the helper, not the son—was up a pole and had connected the wire to make it hot. Gary yelled, "Watch her, Mac. She is hotter than a two-peckered billy goat." The lady went into the house laughing.

Bill T., the fellow who bumped me from leader to line trouble man, was with me when we put the pigs in the backseat. He always used to say, "The first one done is the first one laid off." Another one was, "Buy low, sell high." He always had a knack for that one. One time, Bill loaned me five hundred dollars for two weeks at 10 percent interest, or fifty dollars extra for the two weeks. He counted out ten fifty-dollar bills into my hand, and on the tenth one, as he put it in my hand, he grabbed it back and said, "I'll take the ten percent now." Two weeks later, I gave him the five hundred dollars. He did well for himself before he died. He was still a friend of mine but richer.

Here's one from me. I took the foreman and two linemen flying at lunchtime once in 1971. We were in a Piper Arrow, a complex airplane with retractable gear and over two hundred horsepower. I was getting hours for my commercial pilot license and needed at least two hundred. After about thirty minutes, the foreman got antsy and said, "OK, Bill, we have to land."

I said, "Sorry, but up here, I am the captain." I had to follow the rules to get back to Dayton International Airport. We got back OK and on time, but we never flew at lunchtime again on that crew.

Another time, I took three linemen—Kenny, Deo, and Donnie—flying with me after finishing a job, and we didn't have time for another job. I was the upgraded foreman. 3172R, Donnie. Remember? That was the airplane's number. Donnie always remembered that number for some reason. If we had a problem that day, it wouldn't have been good. That was forty-four years ago. Seems like yesterday. That would have been a hell of an accident report.

CHAPTER 27

Becoming a Flight Instructor

• • •

THE TIME WHEN I WAS becoming a pilot instructor was very difficult, as I had to go through my training while working nights as an emergency lineman. During the night calls, I would arrive at a variety of electrical situations including car wrecks, downed poles and wires, and people who had been killed in one fashion or another. Anything relating to electrical problems, I was there.

While I was an experienced and safe working lineman, one of the things I had to learn when becoming an instructor was how to talk to people. For those who know me, it's hard to believe that I did not know how to talk to people, but it was true. I would talk into a tape recorder and then replay it a lot. Later, as I got more experience as a flight instructor, I had to give these lessons while flying an airplane. Many times, after finishing a lesson plan before a flight, I would play it for my mentor the next day.

One morning, while playing one of my tapes, there was a loud fire truck noise and a siren, and then it was gone. My instructor, Bill Sloan, said, "What the hell was that?"

I replied, "That was a fire truck going to the same call I was going to. It passed me as I was talking into the recorder. There were some twelve-thousand-volt wires on fire."

Bill then said, "You're talking into a tape recorder while speeding down the road to go to a fire with twelve thousand volts in it?"

I replied, "Yes."

He couldn't get over that. He said, "You're a hell of a man." That made me feel so damn good, especially coming from a man who I looked up to so much. He had been a B-26 Martin bomber pilot and had seen a lot. He made me realize that I had also seen a lot. All linemen see a lot; it's part of the job—some good, some bad, and some so funny that you will never forget.

The reason I put this in the book is that my mentor, Bill S., was also a WWII pilot like my friend the general. They both thought that linemen were special people—and they knew about special people, since they were both special. I loved them both.

CHAPTER 28

An Added Story That Just Came to Me, or the Voice

• • •

As I WAS JUST TALKING to my colonel buddy, an F-4 pilot from the 'Nam era, we were discussing loading on a line truck and filing our way out of the parking lot at the power company. I always said that being on a line truck was like going into combat each morning. You never knew what might come up. Later, after becoming an army aero scout in an air cav unit, I realized that was true.

I had the same feeling when I was in a helicopter going out on a training mission in the army guard with just one ship or in a group of three gunships and one scout, which was a popular flight when going after tanks. One day, we found three tanks in the trees because a soldier hung out a white towel to dry. Among green trees, a white towel sticks out like a sore thumb. I spotted them. We went to a holding area for the gunships about a mile away. We took them back to where we found the tanks and simulated taking them out. Later, we had miles' gear for the helicopters and tanks. It was still risky as we flew in the trees twelve feet above them at ninety knots and then slower when we found something. We had a lot of blade strikes and a few chin bubble breaks from hitting the trees. We had the gunships hold in a field as it was safer so as not to get all the helicopters too close together when we were looking for targets. A smaller scout ship was safer going in and out of the trees. The aircraft was an OH58 Bell jet ranger.

We would air taxi out to the runway with the flight and take-off. Sometimes there would be more than one flight going in different directions.

Army aviators also didn't know what might come up. I was never in combat with the national guard, but our training missions were just above the trees, and, at ninety knots, it was a heck of a thrill. As an enlisted aero-scout observer, I was very lucky to be in front of aircraft, and I got a lot of hands-on training with the pilots. Also, many times we were on the range with hot loads of ammunition. That was a hoot, calling for fire and watching the gunships fire at targets just like the helicopter crews going out on training missions.

Line crews filed out of the parking lot, each with a different job to do. At the exit, each truck took to the road with the crew of linemen in the back, going to different destinations and sometimes together to the same one with one crew or many. Being in a line truck was just like taxiing out to the runway in the helicopter or airplanes. We were all going somewhere where there was something that normal people didn't do. Flying in the trees or working twelve thousand volts hot (and, later for me, 345,000 volts hot) is not a normal job.

Sometimes we would hear other crews on the radio calling the dreaded "mayday" when something was wrong somewhere: aircraft down, lineman hurt, electric feeder out, linemen in trouble, or line crew being shot at. DP&L had one lineman who was shot to death near a river. Most of us who did that type of work daily heard it all.

As I am thinking what more to add to this short chapter, I can't help writing something that will *always* keep people safer, no matter what they do: remember the voice.

CHAPTER 29

Pole Buddies

• • •

HERE IS A STORY THAT will let you see the friendship you will form when you work with a fellow a lot, and you keep each other alive while working. In the '60s, we climbed every pole. It was common for poles on main streets of the town to be fifty or sixty feet. Sometimes there were three circuits of primary wire on each pole. Primary wire is high voltage: 2,400 volts was the lowest, and 7,200 volts was the regular primary. Later, 2,400-volt wire was increased to 7,200 volts. When you had to work on the top circuit, you had to climb up through two circuits of hot wire to get up there.

This is where the bucket trucks save time and are so much safer. When we climbed, it took two men to work together to go up. We covered the wire with rubber hoses and rubber blankets. On three-phase circuits, we would put a rubber hose about five feet long on each wire and cover the cross arm that we had to climb over with a rubber blanket around forty-two by forty-two inches or something like that. If it was a big, fat pole, it took two.

The hot crews had two first-class linemen who did that kind of work. When climbing, each lineman checked behind his pole buddy when we moved the wire. When we worked the wire hot, we would stand on an eight-foot ladder with large hooks on it to hang on the cross arms. When there were cross arms three or four feet below where we were, we could just hang the ladder, and the cross arm below would support it. When it was just a single cross arm, we would have a tag line on the ladder to support it. The grunt on the ground would pull it back and tie it to something solid on the ground. Then the linemen would unbelt, step off the pole onto the ladder, and rebelt onto the ladder. Sometimes on big jobs, both linemen would be up there, and both of them would be on hot ladders. Just think of what I have just explained and how long it would take to do this work.

In the buckets, you can go right up to the top wire without even touching the lower wire. Plus, they are so much better on the body. The buckets have made line work much safer.

Now, on to the pole buddies. Being on a hot crew in those days was neat. It was company policy to change crews every eleven to twelve months. So after three or four years, you would be working with four or five different linemen. When you do work like that, you can't help but get close to your pole buddy. You're up there keeping an eye on everything he does, and his eyes are on you. You bond just like you are in the military or a police department. Why wouldn't you do that with someone who is keeping you from being hurt or killed? Yes, we had people killed back then. But you shouldn't brag about the safety record at the present time. It's bad luck to brag about the safety record. But the buckets are much safer.

Giving yourself a pat on the back for being safe is OK. Overly bragging is bad luck. That is what happened the day before we lost our Paul Bunyan of the line department in 1969. Our buddy Duke, who called everyone "little man," was killed along with a brand-new apprentice. Duke had worked for over twenty-four hours straight because there was a storm the night before, and he went back out to a job where a car had hit a pole. Duke was electrocuted changing the pole. That was a blow to all of us at that time; it still chokes me up some just writing about that wonderful man. The sixteen-hour day presently used by many utility companies is safer for the workers. Being overtired is bad. During the Xenia tornado of 1974, we were on the first job for thirty-four hours. That was a tiring day.

I have been retired from that job since 1993, but I am still close to all of my former pole buddies just like I am with the fellows I flew with in the army. It's just human nature to be that way. I have been so lucky not only to still be here but to have been around so many special people. They were crazy guys, and I love them all—even the ones I don't like.

So if some of you end up doing line work, you will have the same kinds of friendships coming your way, even though you won't be climbing as much. Be prepared for getting a new family—you'll see. Also be prepared to get kidded a lot. Remember this. It's what I always told younger people: "People who kid you a lot like you." I said "kid," not "abuse." I don't kid people that I don't like. Good luck to the new linemen, and I know you'll enjoy being in with that kind of group. Damn, you're going to be even luckier: better tools, clothes, money, rest periods, food places that are open all night, company radios in the bucket with you, cell phones also with you, and better information about the job that you're going to. There is just a heck of a lot of things that are better.

Of course, one of the best things that happens when you help people is that you get a great sense of pride for doing a great job. But there is also the response you will get from some grateful people. Some of the fellows met some girls that they dated that way; I was one of them. I am sure that there are a lot of married people in America who met that way. That would be another book of short stories, wouldn't it? Maybe the Hallmark Channel could cover that one too.

This information and all of the stories in this book came from one man. Just think how big this book would have been if they had come from ten linemen. That is the kind of life you will be in for when you join the lineman family. Again, go for it, and have fun. I guarantee that you will. Your joining this elite group of people will also help your country.

Ending Statement

• • •

As I WRITE THIS, WE'RE coming to the end of another winter. This year, winter was pretty calm for most of Ohio, but there were still many storms all over America. The country will soon be hit with the fact that so many of the fine people who are linemen are leaving, mostly due to retirements but for other reasons too. The country needs new linemen for our workforce, which will affect everything the whole country does. Without electricity, the country will come to a standstill.

These jobs are for people who like to work outside and get a good workout while getting paid. Plus, you're helping many people. It is also for the individuals who like some excitement in their lives. Believe me, you will get that, and it is very rewarding to help so many. It will also provide a feeling of camaraderie between you and your fellow workers. That was a special thing we always had and still do. My best friends are the fellows I climbed poles with and army aviators. We still get together quite a lot.

Good luck, and God bless America.

Information on Lineman Schools

• • •

For more information about becoming a lineman, check out this website that lists colleges and authorized programs:

http://www.alexanderpublications.com/pages/lineman-training-in-north-america

Other References

• • •

This book was written by a lineman who is not fancy with words. It is like *Slim*, a famous book about linemen written by William Wister Haines in 1934. It was made into a movie around 1938. Henry Fonda was Slim, and his sidekick was played by Pat O'Brien. Mr. Haines had finished being a lineman and took off awhile to write the book. Later, during WWII, Haines was in the US Army Air Corps in intelligence. After the war, he wrote another great book, *Command Decision*, which was turned into a famous movie with Clark Gable, Van Johnson, and many greater actors. I received this information from Mr. Haines's son, Bill Jr. As with *Slim*, I don't want my book to be changed by the publisher since the way it is written is how linemen talk. The book *Slim* is now being given by unions and companies to workers for doing a good job or for doing something extra and having a customer call in to say that the person did a great job. It's a nice thank-you gift. You can go to this website to order a book if you wish: www.slimthelineman.com.

While I am at it, I would like to mention the "Wichita Lineman," sung by Glen Campbell. It is a great song, and most linemen like it. God bless you, Glen Campbell.

Since I have written this, there has been a new book published called *The American Lineman*, written by Alan Drew, another retired lineman and a teacher at one of the linemen colleges. This book is the Bible of all line work. It has so much information about everything in line work, and it is a great tool for any new person in the field. I bought my copy on Amazon. It would make a wonderful gift for a lineman or a young person looking into that type of work.

About the Author

• • •

William "Bill" H. Arnold Jr. spent thirty-three years as a power lineman for a major power company. He has also been a CFI (certified flight instructor) for twenty-five years and has over six thousand flying hours. He is a retired army aviator, having served in the Ohio Army National Guard.

For more of his stories about line work, see: http://linemanaviator.wordpress.com/.

Bill Arnold in a new Dayton Power & Light jacket

Bill Arnold in army dress blues with his wife, Karen

Made in the USA
San Bernardino, CA
05 December 2016